Focus Open 2011

Internationaler Designpreis Baden-Württemberg

Baden-Württemberg International Design Award

Design Center Stuttgart

avedition

Focus Open 2011
Inhalt

Contents

Dr. Nils Schmid MdL
Stellvertretender Ministerpräsident und Minister für Finanzen und Wirtschaft des Landes Baden-Württemberg

Member of state parliament of Baden-Württemberg Minister of Finance and Industry and Deputy Minister-President

Gutes Design ist weit mehr als nur schönes Aussehen. Die Vorteile von Design liegen auf der Hand: Design macht neugierig und wirkt identitätsstiftend. Gutes Design bedeutet unter anderem Innovation, Nachhaltigkeit und Benutzerfreundlichkeit.

Good design is far more than attractive appearance. The benefits of design are obvious: design arouses curiosity and creates identity. Good design means things such as innovation, sustainability and user-friendliness.

Der Focus Open mit den Preisen »Focus in Gold« und »Focus in Silber« hat seinen festen und anerkannten Platz, wenn es um die Auszeichnung von herausragendem Design geht. Renommierte Unternehmen aus Baden-Württemberg, Deutschland und Europa bewerben sich bei Focus Open.

Wie bei einem internationalen Designwettbewerb zu erwarten, sind auch 2011 unter den Preisträgern Unternehmen aus dem Ausland vertreten. Als Minister für Finanzen und Wirtschaft des Landes Baden-Württemberg freue ich mich über die internationale Resonanz und auch darüber, dass zahlreiche heimische Unternehmen mit ihren Produkten beim diesjährigen Focus Open Wettbewerb prämiert wurden.

Allen Gewinnern möchte ich herzlich gratulieren. Jedem Preisträger wünsche ich einen maximalen Nutzen durch die Auszeichnung mit dem Internationalen Designpreis Baden-Württemberg, zum Beispiel durch hohe internationale Aufmerksamkeit im Rahmen der Preisverleihung, der Ausstellung, der Internetpräsenz sowie der weiteren Präsentation ihrer Produkte in dieser Publikation. Der Focus Open hilft seinen Preisträgern, indem sie mit der Auszeichnung für exzellente Produktqualität werben können und unterstützt sie wirkungsvoll in ihrer Marktkommunikation.

Professionelles Design, wie es hier in diesem Buch gezeigt wird, stellt einen Mehrwert für jedes Unternehmen dar. »Das Design sollte das Produkt sozusagen zum Sprechen bringen«, so hat Dieter Rams, deutscher Industriedesigner der Moderne, die Bedeutung von Design in einer Metapher gewürdigt. Gutes Design ist weit mehr als nur schönes Aussehen. Die Vorteile von Design liegen auf der Hand: Design macht neugierig und wirkt

With its »Focus in Gold« and »Focus in Silver« awards, Focus Open has a firm place as a recognized arbiter of excellent design. Renowned companies from Baden-Württemberg, Germany and Europe enter their products for the Focus Open competition.

And as is to be expected of an international design competition, the 2011 award-winners again include companies from outside Germany. As the Baden-Württemberg Minister for Finance and Industry, I am pleased by this international response, and also that many local companies have been singled out for awards in this year's Focus Open competition.

I would like to offer my warmest congratulations to all the award-winners. I hope that each award-winner will get the maximum benefit from being selected for the Baden-Württemberg International Design Award – for example, as a result of heightened international attention in the course of the award-giving ceremony, the exhibition, publicity on the web, or the presentation of their products in this yearbook. The benefit of Focus Open for its award-winners is that they can use the award to advertise the excellent quality of their products, as well as to effectively support their marketing communication.

As this publication shows, professional design represents value added for any company. When asked what the significance of design was, the modern German industrial designer Dieter Rams said: »Design should make the product speak, as it were«. Good design is far more than attractive appearance. The benefits of design are obvious: design arouses curiosity and creates identity. Good design means things such as innovation, sustainability and user-friendliness. It is something to

Dr. Nils Schmid MdL
Stellvertretender Ministerpräsident und Minister für Finanzen und Wirtschaft des Landes Baden-Württemberg

Dr Nils Schmid
Member of state parliament of Baden-Württemberg Minister of Finance and Industry and Deputy Minister-President

identitätsstiftend. Gutes Design bedeutet unter anderem Innovation, Nachhaltigkeit und Benutzerfreundlichkeit. Hierfür stehen auch große Designer aus Baden-Württemberg wie Wilhelm Wagenfeld, Otl Aicher und Kurt Weidemann sowie »Produktklassiker« wie der Lamy 2000, die Gardena Schlauchkupplungen oder der Bosch Akkuschrauber IXO.

In Baden-Württemberg wird professionelles Design durch das Design Center Stuttgart kompetent gefördert. Die Basis für das Design Center Stuttgart legte eine Mustersammlung im Haus der Wirtschaft vor über 100 Jahren. Aus aller Welt wurden Gebrauchsgüter als Vorbilder für die hiesige Wirtschaft ausgestellt. Diese Produkte wurden von Weltausstellungen »mitgebracht«. Heute fördert das vom Land getragene Design Center Stuttgart gutes Design mit Fachtagungen wie zum Beispiel Cross Culture Branding und Ausstellungen, sondiert das internationale Marktgeschehen und gibt Antworten zu aktuellen Anforderungen an Designentwicklung und Markenstrategie im internationalen Wettbewerb.

Deshalb ist das Design Center Stuttgart ein wichtiger Kooperationspartner im neuen, landesweiten Netzwerk »Cluster Kreativwirtschaft Baden-Württemberg«, welches 2011 seine Aktivitäten aufgenommen hat. Das Netzwerk wird vom Ministerium für Finanzen und Wirtschaft mit europäischen Geldern in Höhe von 500.000 Euro gefördert. Ziel ist es, die vielfältigen Teilbranchen der baden-württembergischen Kreativ- und IT-Wirtschaft zu unterstützen und noch besser miteinander zu vernetzen. Im Sommer 2011 startete das Cluster Kreativwirtschaft in einer Veranstaltung »Marktplatz für Mittelstand und Kreative« im Stuttgarter Rathaus mit einem Schwerpunkt auf der Designwirtschaft.

Förderung der Designwirtschaft ist Wirtschaftsförderung. Deshalb wird sich die neue Landesregierung Baden-Württemberg auch in Zukunft des Themas Design unterstützend annehmen.

which great designers from Baden-Württemberg have contributed, such as Wilhelm Wagenfeld, Otl Aicher and Kurt Weidemann, as well as classic products from the state, such as the Lamy 2000, Gardena hose connectors or the Bosch Ixo cordless screwdriver.

In Baden-Württemberg, professional design is professionally promoted by the Design Center Stuttgart. The roots of the Design Center Stuttgart go back more than 100 years, to a collection of patterns in the »Haus der Wirtschaft«. Consumer goods from all over the world were exhibited as models for local industry. The products were brought to Stuttgart from world expos. Today, the state-funded Design Center Stuttgart promotes good design with specialist conferences on topics such as cross-cultural branding and with exhibitions. It also keeps its finger on the pulse of international markets and gives answers to topical challenges to design development as well as advice about brand strategy in an international competitive environment.

This is why the Design Center Stuttgart is an important cooperation partner in the new, state-wide »Baden-Württemberg creative industry network«, which came into being in 2011. The Finance and Industry Ministry supports this network with 500,000 euros from EU funds. The network's aim is to support the many individual segments of the creative and IT industry in Baden-Württemberg, and to help them to network better. In the summer of 2011, the creative industry cluster began its work with a »marketplace for SMEs and creatives« event at Stuttgart city hall – an event that focused on the design business.

Promoting the design business is promoting industry. For this reason, the new state government of Baden-Württemberg will continue to support design in the future as well.

Focus Open
2011

Tendenzen aufzeigen, innovative Lösungen aufspüren und Maßstäbe für Produktqualität setzen.

Highlighting trends, uncovering innovative solutions and setting standards for product quality.

Johannes Schmalzl
Regierungspräsident des Regierungsbezirkes Stuttgart
Chairman of the Stuttgart Regional Council

Designer und Unternehmen aus aller Welt reichten auch 2011 wieder ihre innovativsten Produkt- und Konzeptlösungen beim Focus Open, dem Internationalen Designpreis Baden-Württemberg ein und stellten sich dem Wettbewerb für herausragendes und zukunftsfähiges Design.

Die Besten – die Preisträger des »Focus in Silber« für hervorragende Designlösungen und des »Focus in Gold« für ausgezeichnete, exzellente Gestaltung – werden in diesem Jahrbuch präsentiert.

Insgesamt wurden 357 Produkte aus Deutschland, Österreich, Schweiz, Italien, Spanien, Dänemark, Niederlande, Belgien, Großbritannien, Frankreich und den USA eingereicht.

Die Teilnehmer konnten ihre Einreichungen folgenden Kategorien zuordnen:
— Produktion, Montage, Logistik
— Medizin, Rehabilitation, Geriatrie
— Bad, Sanitär, Wellness
— Küche, Haushalt
— Tisch- und Küchenkultur
— Wohnen
— Objekt
— Ambiente, Lifestyle
— Beleuchtung
— Kommunikation, Audio / Video, Optik
— Freizeit, Outdoor
— Sport
— Kids, Familie
— Architektur
— Public Design
— Transport, Verkehr

Im Zuge ihrer Beurteilung definierte die Jury die zusätzliche Kategorie: Konzept, Nachhaltigkeit.

Die Herausforderung der Bewertung
357 Produkte aus 17 Kategorien in zwei Tagen – eine Herausforderung für jede ernsthaft arbeitende Jury.

Um wirklich jedem Produkt eine Chance zu geben, wurde keine Vorauswahl getroffen. Jede einzelne Idee wurde intensiv begutachtet, erläutert, diskutiert und bewertet. Die Jurierung umfasste dabei mehrere Durchgänge. Diese intensive Auseinandersetzung mit den Einreichungen führte, gerade bei besonders innovativen, nicht »gelernten« Produkten und Konzepten, zu lebhaften, zum Teil durchaus kontroversen aber immer konstruktiven Diskussionen. Die interdisziplinäre Zusammensetzung der Jury mit Erfahrungen aus verschiedensten Branchen und Produktkate-

In 2011, designers and companies once more submitted their most innovative products and ideas for Focus Open, the Baden-Württemberg International Design Award, and competed for this award for outstanding design with a future.

The best of these entries – the winners of the »Focus in Silver« for excellent design solutions and of the »Focus in Gold« for outstanding, superb design – are presented in this yearbook.

The products submitted for judging came from Germany, Austria, Switzerland, Italy, Spain, Denmark, the Netherlands, Belgium, the UK, France and the United States.

Participants could assign their entries to one of the following categories:
— Production, assembly, logistics
— Medicine, rehabilitation, geriatric medicine
— Bathrooms, sanitary installations, wellness
— Kitchen, household
— Tableware and kitchenware
— Living
— Contract furniture
— Ambience, lifestyle
— Lighting
— Communication, audio / video, optical equipment
— Leisure, outdoor
— Sport
— Children, family
— Architecture
— Public design
— Transport, traffic

In the course of their assessment, the judges defined an additional category:
Design studies for sustainability

The challenge of assessing the entries
357 products in 17 categories in 2 days – a challenge for any respectable judging panel.

To give each product a real chance, there was no pre-selection. Each single idea was intensively appraised, explained, discussed and judged. Judging went through several stages. Especially in the case of products and concepts that were particularly innovative and unprecedented, this intensive critical ex-

Christiane Nicolaus
Leiterin des Design Center Stuttgart
Head of Design Center Stuttgart

gorien trug zu fairen und kompetenten Bewertungen aus den unterschiedlichsten Perspektiven bei.

Die Experten der Jury legten für ihre Bewertung folgende Kriterien zu Grunde:
— Gestaltungsqualität
— Funktionalität
— Innovationsgehalt
— Ergonomie, Interfacedesign
— Produktgrafik beziehungsweise -typografie
— Ökologie, Nachhaltigkeit
— Wertigkeit, Anmutung
— Emotionalität
— Markenkonformität

Die Besten
Die Jury vergab 2011 insgesamt 91 Auszeichnungen. Der »Focus in Gold« wurde elf Mal vergeben, 80 weitere Produkte erhielten den »Focus in Silber«.

Fazit
Focus Open, der Internationale Designpreis Baden-Württemberg, ist seit vielen Jahren fester Bestandteil der Wettbewerbslandschaft in Deutschland. Er ist einer der wenigen nicht kommerziellen Wettbewerbe, dessen Intention das Aufzeigen von Tendenzen im Design, das Aufspüren innovativer Lösungen und in erster Linie die Förderung der jährlichen Preisträger ist.

Der renommierte Designpreis setzt Maßstäbe für Produktqualität und unterstützt gleichzeitig Unternehmen und Designstudios mit den Auszeichnungen »Focus in Gold« und »Focus in Silber« bei der Marktkommunikation.

Die Wertschätzung gegenüber den Teilnehmern und der persönliche Umgang mit den Preisträgern, von der ersten Benachrichtigung bis zur Preisverleihung, verleiht diesem Wettbewerb Charakter und macht ihn, in der heutigen Flut von Wettbewerben, zu etwas ganz Besonderem.

Gerade durch die Aspekte der Seriosität und der nichtkommerziellen Ausrichtung hat der Designpreis Baden-Württemberg im Laufe der Jahre im In- und Ausland hohes Ansehen und seitens der Wirtschaft und Fachwelt hohe Wertschätzung erfahren.

amination of the submissions led to lively discussion, which could sometimes be controversial but was at all times constructive. The interdisciplinary make-up of the judges panel, with their experience from a diverse range of industries and product categories, contributed to assessments that were fair and competent, and took account of various perspectives.

The judges based their assessment on the following criteria:
— Design quality
— Functionality
— Innovative content
— Ergonomics, interface design
— Product graphics and typography
— Ecology, sustainability
— Quality, grace
— Emotional appeal
— Brand conformity

The best
The judges awarded a total of 91 prizes in 2011. The »Focus in Gold« was awarded 11 times, and the »Focus in Silver« was conferred on a further 80.

Conclusion
Focus Open, the Baden-Württemberg International Design Award, has been a firm part of the German competition scene for many years now. It is one of the few non-commercial (state) competitions whose intention is to highlight design trends, uncover innovative solutions and, first and foremost, to promote its annual award-winners.

This renowned design award sets standards for product quality and, at the same time, its Focus in Gold and Focus in Silver awards support companies and design studios in their marketing communication.

The appreciation shown for participants and the personal contacts with the award-winners, from the initial notification to the presentation of the award, is what gives this competition its character. It makes it something special in today's plethora of competitions.

It is especially the competition's respectability and non-commercial orientation that have given the Baden-Württemberg design award a good reputation both in Germany and beyond its borders over the years, and that ensure that it is held in high esteem by business and the design profession alike.

Dr. Nicole Busch
Buschwerk, Hamburg

Sie ist ein Instrument, um Designern oder Produktentwicklern Innovationsmöglichkeiten und Inspiration an die Hand zu geben. Bei der User Experience Research geht es darum, herauszufinden, welche Bedürfnisse, Wünsche oder Probleme Menschen bei Produkten und Services haben. Es gibt nichts, worauf man sie nicht anwenden kann.

It is a tool for providing designers and product developers with inspiration and possibilities for innovation. User experience research is about finding out what needs, wishes or problems people have with products and services. There is nothing to which it cannot be applied.

Bei meiner Arbeit, der User Experience Research, geht es darum, herauszufinden, welche Bedürfnisse, Wünsche oder Probleme Menschen bei Produkten und Services haben, um dem Design oder der Produktentwicklung Innovationsmöglichkeiten und Inspiration an die Hand zu geben. Es gibt fast 100 Methoden, um Informationen zu gewinnen, was Menschen sagen und tun, aber auch denken und fühlen. Man kann beobachten, befragen, Probanden selbst Dinge ausprobieren oder Prototypen bauen lassen, ethnografische Interviews führen, um Daten aus unterschiedlichen Kulturen oder dortige Anwendungsweisen zu erhalten. Man schafft damit Grundlagen für die Designentwicklung und bekommt Inspirationen, die man vorher nicht für möglich erachtet hätte.

Für eine Sportfirma habe ich Sportler beim Laufen begleitet und geschaut, welche Tools sie dabei benutzen, warum sie laufen, wie sie ihre Daten verwenden. Es zeigte sich, dass sie eine Art persönlichen Trainer wollen, der sie begleitet oder auch Trainingspläne erstellt, abgestimmt auf ihre Motivation oder ihr Können. Auf Basis dieser Studien wurde so eine Gerätefamilie entwickelt. Dafür wird ein hervorragender Service geboten, eine wichtige Anforderung, die sich aus den Untersuchungen ergeben hat. Auch damit kann man sich deutlich von der Konkurrenz unterscheiden.

Buschwerk, Hamburg

The work I do is in user experience research. This is about finding out what needs, wishes or problems people have with products and services, and then to make this information available in the design or product development process as an inspiration or possibility for innovation. There are nearly 100 ways of gaining information about what people say and do, think and feel. You can observe and ask, get test persons to try things out or build prototypes, or conduct ethnographic interviews in order to gather data from different cultures about the way they use products. In this way, you form a basis for design development and get inspiration you would previously never have thought possible.

For a sports goods company, I have accompanied athletes on runs and looked at the kind of tools they use, why they run, and what they do with their data. What emerged was that they want a kind of personal trainer to accompany them and draw up training plans for them that are tailored to their motivation and ability. These studies then formed the basis

Nach Banklehre und Biologiestudium arbeitete Dr. Nicole Busch erst bei Designaffairs als Ergonomiespezialistin, dann bei IDEO als Human Factors Researcher in München, Palo Alto, San Francisco, Chicago, Boston und London und als Projektmanagerin beim Trendbüro, Hamburg. Parallel dazu promovierte sie im Design Engineering an der TU Delft mit Schwerpunkt angewandte Ergonomie und Design. Seit 2007 ist sie selbstständig als Expertin für User Experience Research und Design. Recherche und Ideenentwicklung für Produkt- und Servicedesign, Studien und Workshops für Unternehmen wie Adidas, Design 3, Xing sind Teil ihres Repertoires. Ihre Themen sind der User, Design und Usability u.a. in den Bereichen Konsumgüter, Haushaltsgeräte, Medien, Sport, Medizintechnik und Mobilität. Zusätzlich lehrt sie Ergonomie an der Burg Giebichenstein Kunsthochschule Halle und an der Muthesius Kunsthochschule Kiel die Methoden des Human Centred Design Research.

Following an apprenticeship in a bank and university studies in biology, Dr. Nicole Busch worked first of all as an ergonomics specialist with designaffairs, then with IDEO as a human factors researcher in Munich, Palo Alto, San Francisco, Chicago, Boston and London. She has also been a project manager at Trendbüro in Hamburg. During this time, she wrote a PhD in design engineering at Delft university, focusing on applied ergonomics and design. Since 2007, she has been a freelance expert on user experience research and design. Her repertoire includes research and developing ideas for product and service design, as well as studies and workshops for companies such as Adidas, Design 3 and Xing. Her focus is on users, design and usability in areas such as consumer goods, household appliances, media, sport, medical technology and mobility. She also teaches ergonomics at the Burg Giebichstein College of Art in Halle, and methods of human-centred design research at the Muthesius College of Art in Kiel.

Es sind immer verschiedene Menschen, die an einem Produkt- oder Service-Prozess beteiligt sind, am Beispiel einer Heizung: der Kunde, der sie nutzt und sich wärmt, der Ingenieur, der sie nach neuesten Richtlinien für Energieeffizienz entwickelt, der Monteur, der sie wartet, der Schornsteinfeger, der sie abnimmt und den Schlot reinigt. Alle sind Stakeholder in einem bestimmten Prozess. Im Medizinbereich sind es meist besonders viele: der Arzt, die Schwester, der Patient, der Klinikchef, der einkauft, der Techniker, der das Gerät anschließt. Alle sollten bei der Produkt- oder Servicegestaltung bedacht werden.

Es gibt nichts, worauf man User Experience Research nicht anwenden kann. Jede Industrie bietet großes Potenzial, den Menschen und Kunden als Inspirations- und Innovationsquelle zu nutzen.

for a family of sports equipment. There is also an excellent service to go with them. This was an important requirement to emerge from our studies. It is also a way of standing out clearly from the competition.

There are always different people involved in a product or service process. Take the example of a heating system: there is the customer who uses it to get warm, the engineer who develops it to comply with the latest energy-efficiency regulations, the technician who services it, and the chimney sweep who inspects it and cleans the flue. They are all stakeholders in a certain process. In the medical sphere, there is usually an especially large number of stakeholders: doctor, nurse, patient, the hospital director purchasing appliances, and the technician connecting them. All of them should be considered when designing the product or service.

There is nothing to which user experience research cannot be applied. Every industry has great potential for using people and customers as a source of inspiration and innovation.

Roland de Fries
Kuypers Branding, New York

Kuypers Branding, New York

Der Trend geht vom klassischen Designumfeld hin zu Digital- und Mobilitätsaspekten: Steuerung einer Heizungsanlage über ein Benutzerinterface von Smartphones, aufblasbare modulare Räume, mit denen man Erlebniswelten schaffen kann, oder »Smart«-Studien, bei denen das iPhone zum zentralen Instrument für die Fahrzeuge wird. Wird hier Gestaltung neu definiert?

The trend is moving away from the classic design environment to digital, mobile worlds: controlling a heating system via a smartphone user interface, inflatable modular rooms that can be used to create atmospheric spaces, »smart« studies in which an iPhone is the central instrument in the vehicles. Is design being re-defined here?

Was ich spannend fand, waren einige gelungene Produkte mit Steuerung über ein Benutzerinterface von Smartphones oder iPhone, zum Beispiel die Heizungsanlage von Junkers, die man mit einer iPhone-App ortsunabhängig ansteuern kann oder die »Smart«-Studien, bei denen das iPhone zum zentralen Instrument für die Fahrzeuge wird. Es ist einiges im Wandel, alles wird zunehmend mobiler. Man merkt inzwischen, dass diese Smartphone-Apps Sinn machen und keine Spielereien mehr, sondern funktional durchdacht sind.

Der Trend geht vom klassischen Designumfeld hin zu Digital- und Mobilitätsaspekten. Dazu gehören auch die aufblasbaren modularen Räume, mit denen man Erlebniswelten schaffen kann und sich nicht mehr Gedanken über Gewicht, Lagerung und Transport machen muss. Man faltet sie klein zusammen, wenn man sie braucht, blasen sie sich einfach auf. Eine neue Definition von Gestaltung.

Ich denke, dass bei einem Smartphone oder iPad ein intuitiver Einstieg bereits vorhanden ist, bevor man mit der eigentlichen Anwendung in Kontakt kommt. Das erleichtert den Zugang, fordert aber den Designer heraus: Wie kann er das Produkt visuell abheben, um zu verdeutlichen, dass es sich nicht lediglich um eine normale iPhone-App, sondern einen eigenständigen Markenauftritt handelt und gleichzeitig die intuitiven und funktionalen Kriterien erfüllt?

What was exciting for me was a number of well executed products controlled using a smartphone or iPhone interface. For example, the Junkers heating system that can be controlled from anywhere using an iPhone app, or the »smart« studies in which an iPhone is a central instrument in the vehicles. A lot of changes are happening here; everything is becoming increasingly mobile. It is now becoming clear that these smartphone apps are also meaningful, not just fun and games. They are functionally very well designed.

The trend is moving away from the classic design environment to digital, mobile worlds. This also includes inflatable, modular rooms that can be used to create atmospheric spaces, without having to worry about weight, storage and transport. They are simply folded up small and inflated again when they are needed. In fact, this is a new definition of design.

For me, an intuitive knowledge already exists in a smartphone or iPad before you even come into contact with the actual applica-

Der Kommunikationsdesigner und Branding Experte aus Stuttgart war in den 1990er Jahren selbstständig für namhafte Unternehmen in Deutschland tätig, wie Porsche, Breuninger und Deka Bank, bevor Roland de Fries den Sprung in die USA wagte. In New York arbeitete er zunächst als Design Director bei Ogilvy für Firmen wie IBM, Unilever und Bulgari, bevor er mit Kuypers Branding seine eigene Agentur gründete. Die Kundenliste umfasst Blue Chips wie American Express, den französischen Flugzeugbauer Dassault Falcon Jet, Time Warner Center, Quontic Bank sowie eine Reihe deutscher und schweizer Lifestyle-Marken. Jüngster Spross ist seine Gründung der US-basierten Handelsgesellschaft »Hudson Vandam LLC«, die speziell mittelständische Unternehmen und Marken beim erfolgreichen Einstieg und Vertrieb auf dem nordamerikanischen Markt unterstützt.

Originally from Stuttgart, the communications designer and branding expert Roland de Fries worked for renowned companies such as Porsche, Breuninger, and Deka Bank in the 1990s before moving to the United States. There he started out as design director at Ogilvy in New York City, working for companies such as IBM, Unilever and Bulgari. He then established his own branding agency, Kuypers Branding. Its impressive client list includes blue-chip companies such as American Express, the French aviation company Dassault Falcon Jet, Time Warner Center and Quontic Bank, as well as a number of German and Swiss lifestyle brands. Recently he established his own U.S.-based distribution company »Hudson Vandam LLC«, which focuses on helping German medium-size companies brands and products to become successfully established in the North American market.

»Made in Germany« steht in den USA für Qualität und Bauhausprodukte. Jetzt kommt ein neuer Trend hinzu: Gutes deutsches Ingenieurswesen trifft über Interface und Technologie amerikanische Marketingphilosophie: Ausgesprochen Consumer- und User-orientiert in den USA versus sehr vom Ingenieur und Produkt her gedacht in Deutschland! Das finde ich spannend, weil dadurch das Image »Made in Germany« am amerikanischen Markt aufgefrischt werden kann. Die USA sind ja für Deutschland immer noch der stärkste Exportmarkt außerhalb der EU.

Ein anderer Trend sind erneuerbare Energien. Da ist Deutschland führend und die USA sind noch nicht so weit: ein Riesenmarkt für deutsche Hersteller. Das wird ebenfalls spannend.

Der Focus-Designpreis ist dabei ein gutes Bindeglied, bei dem jüngste Entwicklungen und überzeugende Produkte noch mehr in die Öffentlichkeit katapultiert werden.

tion. This makes use easier, but also challenges the designer. How can designers set the product apart visually, so that it is not just some ordinary iPhone app but a product that stands alone in the market, and this while fulfilling intuitive and functional criteria?

In the U.S., »made in Germany« stands for quality and Bauhaus products. Now this is being joined by a new trend: a marriage of good German engineering and American marketing philosophy, which come together on the level of interface and technology. We have a marked consumer and user orientation in the U.S., while in Germany the orientation is very much towards the engineer and the product. For me, this is exciting, because it can rejuvenate the image of »made in Germany« in the U.S. market. For Germany, after all, the United States is still the biggest export market after the EU.

Renewable energy is another trend. Germany leads the way here, with the U.S. still lagging behind. This is a huge market for German manufacturers. This will also be exciting.

The Focus design award is a good go-between here, bringing the latest developments and outstanding products even more effectively into the public eye.

Stefan Grobe
Defortec GmbH, Dettenhausen

Defortec GmbH, Dettenhausen

Gerade im Investitionsgüterbereich ist eine neue Designaffinität und Begeisterungsfähigkeit der Kunden zu spüren, beispielsweise auf dem Gebiet der Produktionstechnologien von erneuerbarer Energie. Viele dieser dynamischen Unternehmen mit jungen Vorständen erkennen die Chance, die Design bietet, anders zu sein und anders zu denken und setzen diese Chance in aller Konsequenz um.

Especially in capital goods, we notice a new affinity and enthusiasm for design among clients, for example in the field of production technologies for renewable energies. Many of these dynamic companies with their young managers understand the opportunity that design offers to be different and think differently, and are eagerly grasping this opportunity.

Die besondere Faszination des Designs in der heutigen Zeit ist die Durchdringung der Produktwelten in einer neuen Tiefendimension. Design bleibt nicht mehr an der Oberfläche, sondern verändert die Produkte grundsätzlich in ihrer Funktionalität, ihrem Gebrauchswert (Usability) und Emotionalität. Design sucht nach Möglichkeiten, um die herausragende Eigenschaft des Produktes für den Bediener erlebbar zu machen.

Gerade im Investitionsgüterbereich spüren wir eine neue Designaffinität und Begeisterungsfähigkeit der Kunden. Wir arbeiten im Augenblick verstärkt im Bereich der Produktionstechnologien von erneuerbarer Energie. Viele dieser dynamischen Unternehmen mit jungen Vorständen erkennen die Chance, die Design bietet, anders zu sein und anders zu denken und setzen diese Chance in aller Konsequenz um.

Gleichzeitig fordern sie vom Entwickler ein höchstes Maß an innovativem Denken, und zwar in Bezug auf deren wirtschaftliche Umsetzbarkeit und einen frühen Return on Investment. Gelingt diese Kombination aus »Design to Cost« und einer neuen Produktemotionalität, die technisch komplexe Lösungen erlebbar macht, entstehen nachhaltig erfolgreiche Produkte.

What is especially fascinating about today's design is the way it is permeating product worlds more profoundly than before. Design is no longer staying on the surface, but is fundamentally changing products in their functionality, usability and emotionality. Design is looking for ways of making a product's outstanding properties something the user can grasp.

Especially in capital goods, we notice a new affinity and enthusiasm for design among clients. At the moment, we are especially involved in production technologies in the renewable energy field. Many of these dynamic companies with their young managers understand the opportunity that design offers to be different and think differently, and are eagerly grasping this opportunity.

At the same time, they demand the maximum of innovative ideas from their engineers, with respect to both economic feasibility and early return on investment. If they

Stefan Grobe absolvierte 1994 mit Auszeichnung sein Industrial Design Studium an der Universität Essen, Folkwang Hochschule. Nach Tätigkeiten in renommierten Designbüros führte er 13 Jahre lang erfolgreich als Mitinhaber das Stuttgarter Büro BGP Design und gewann zahlreiche Auszeichnungen und Preise. 2011 gründete er die Defortec GmbH, die intelligente Designlösungen für komplexe technische Systeme entwickelt. Schwerpunkte bilden regenerative Energien, Kunststoffverarbeitung sowie Labor- und Medizingerätetechnik mit Kunden wie Centrotherm, Vestas, Ferromatik Milacron und GP Solar. User Interface Entwicklungen und Produktvisualisierungen runden das Leistungsspektrum ab. Stefan Grobe hält Lehraufträge an der HFG Schwäbisch Gmünd zum Thema Produktgestaltung und ist gefragter Referent auf Veranstaltungen zum Thema Produktentwicklung und Innovation.

Stefan Grobe graduated in industrial design from the University of Essen-Folkwang in 1994. After working in renowned design studios, he successfully managed the Stuttgart-based studio BGP Design for 13 years. As its co-owner, he won numerous awards and prizes. In 2011, he set up Defortec GmbH, which develops smart design solutions for complex technical systems. Its focus is on renewable energy, plastics processing and laboratory and medical apparatus, with clients such as Centrotherm, Vestas, Ferromatik Milacron and GP Solar. The studio's range of work is rounded off by user-interface development and product visualization. Stefan Grobe lectures on product design at the Schwäbisch-Gmünd University of Design. He is a popular speaker at events on product development and innovation.

Industriedesign wird dabei zunehmend international und ist eingebettet in eine globalisierte Entwicklungs- und Produktionskette. Der Designer muss verschiedene Unternehmenskulturen und Besonderheiten der Produktionsstandorte berücksichtigen – bis hin zum verschiedenen Farbempfinden der Kundenmärkte.

Das Anforderungsprofil an Designer erweitert sich also ständig. Sie müssen mehr sein als Gestalter, sie werden zum Suchenden, Entdecker und Finder.

succeed in this combination of design to cost and new product emotionality that makes technically complex solutions tangible, then lastingly successful products will arise.

Industrial design is becoming increasingly international, and is embedded in a globalized chain of development and production. The designer has to consider different corporate cultures and the peculiarities of production locations – right down to different perceptions of colour by different markets.

In other words, more and more demands are being made of designers. They have to be more than just creators of form. They are becoming searchers, discoverers and finders.

Markus Jehs
Jehs&Laub, Stuttgart

Jehs&Laub, Stuttgart

Wichtig ist die Kunst des Weglassens. Weg vom Übergestalteten hin zum Authentischen und Reduzierten, zur Verwendung von naturbelassenen oder recycelten Materialien. Bei Beginn eines Projektes stellen sich immer die gleichen Fragen: Warum macht man das? Braucht man das überhaupt? Dass es »dem Fortschritt dient«, ist die Grundvoraussetzung für jede neue Produktentwicklung.

The art of doing without is important. It's good to move away from the over-designed to the authentic and reduced, to the use of untreated and recycled materials. At the beginning of a project, it is always the same questions that are asked: Why do we do this? Do we even need this? The idea of serving progress is the precondition of any new product development.

Ein wesentlicher Punkt bei der Auswahl der Produkte ist für mich, welche Strategien die Firmen haben und wie sie auf bestimmte Entwicklungen reagieren. Dass ein konventioneller Autohersteller, wie Smart, über Alternativen bei den Antriebstechnologien sowie bei der Fahrzeugrange, also auch zweirädrige Fahrzeuge, wie Motorroller oder Elektrofahrräder, nachdenkt, finde ich gut. Denn als Juror kann man, wenn man so etwas prämiert, auch Zeichen setzen.

Wichtig finde ich auch die Kunst des Weglassens. Weg vom Übergestalteten, Übertriebenen, hin zum Authentischen und Reduzierten, zur Verwendung von naturbelassenen oder recycelten Materialien wie zum Beispiel bei dem Zelt Blue One von Vaude. Wenn wir ein Projekt beginnen, stellen wir uns immer den gleichen Fragen: Warum machen wir das? Muss das sein? Braucht man das überhaupt? Wenn wir etwas sinnlos finden, dann machen wir es auch nicht. Die Vorstellung, dass es »dem Fortschritt dient«, ist die Grundvoraussetzung für jede neue Produktentwicklung.

Wir arbeiten im Bereich Möbel oder Objektmöbel, aber auch Tableware mit weltweit führenden Firmen, die mit Design groß geworden sind. Wenn wir unsere Ideen präsentieren, dann heißt es nicht gleich »geht nicht«.

For me, one of the most important points when selecting products is what strategies companies have and how they react to certain developments. I like the way a conventional carmaker such as smart thinks about alternative powertrains and an alternative range of vehicles, including two-wheelers such as scooters and pedelecs. As a judge, you can also send out a signal when you confer an award on such a product.

I also think that the art of doing without is important. It's good to move away from the over-designed and exaggerated to the authentic and reduced, to the use of untreated and recycled materials – in Vaude's Blue One tent, for example. Whenever we start a project, we always ask ourselves the same questions: Why do we do this? Do we have to? Do we even need this? It we don't think something makes sense, then we don't do it. The idea of serving progress is the precondition of any new product development.

Markus Jehs zählt zu den gefragtesten deutschen Produktdesignern bei weltweit agierenden Markenunternehmen. Nach seinem Studium an der Hochschule für Gestaltung in Schwäbisch Gmünd gründete er 1994 mit Jürgen Laub ein gemeinsames Büro in Stuttgart. Heute stehen namhafte Hersteller aus der Einrichtungsbranche wie Cassina, Cor, Fritz Hansen, Knoll, Stelton und Wilkhahn auf der Kundenliste des schwäbischen Erfolgsduos. Ihr Design vereint »typisch deutsche« Eigenschaften wie Funktionalität und Klarheit mit außergewöhnlichen Konzepten und cleverem Materialeinsatz. Bei ihren Stühlen, Tischen, Sofas und Leuchten gehen Logik und Fantasie widerspruchsfrei zusammen. Neben Einzelprodukten entwickeln sie auch komplette Raumkonzepte wie etwa die weltweiten Showrooms für Mercedes-Benz oder eine Suite im nordschwedischen Icehotel.

For globally operating brand names, Markus Jehs is one of the most sought-after German product designers. In 1994, after studying at the Schwäbisch-Gmünd University of Design, he founded a studio together with Jürgen Laub in Stuttgart. Today, this successful duo's clients include renowned manufacturers from the interior decoration sector, such as Cassina, Cor, Fritz Hansen, Knoll, Stelton and Wilkhahn. Their design combines typical German qualities such as functionality and clarity with unusual concepts and a clever use of materials. Their chairs, tables, sofas and luminaries harmoniously combine logic and fantasy. Apart from single products, they also develop complete room concepts, such as the global showrooms of Mercedes-Benz or a suite in the northern Swedish Icehotel.

Unsere Kunden verstehen, was wir meinen, und versuchen alles, um diese Ideen umzusetzen. Man arbeitet dort mit den besten Leuten, die nicht kurz vor dem Ziel aufgeben, sondern wissen, dass die letzten fünf Prozent die entscheidenden sind.

Oft ist unser Antrieb der, dass wir Dinge machen, die wir uns selbst wünschen. Damit ist der Anspruch hoch und man identifiziert sich automatisch mit dem Endkunden. Genauso wie ein Schauspieler, der, wenn er die Rolle nicht komplett annimmt, sie auch nicht überzeugend spielen kann.

In the area of (contract) furniture, or also of tableware, we work together with globally leading companies that grew up with design. When we present our ideas, there is never a knee-jerk »no way« reaction. Our customers understand what we mean, and do all they can to put our ideas into practice. We work there with the best people, people who don't give up just before the finishing line, but who know that the final five percent is what counts.

We are often driven by the fact that we make things we ourselves want to have. This means that the standards we set are high, and that we automatically identify with the end-customer. In this way, we're just like actors. An actor cannot play a role convincingly unless he fully accepts his role.

Christoph Daniel Jia
Linya Int'l Strategic Foresight,
Shanghai

In China hat »Made in Germany« eine unglaubliche Strahlkraft. Die Chinesen haben eine große Sehnsucht nach Authentizität, die sie in deutschen Produkten finden. Dort denkt man eher an den Nutzer und die Bedienbarkeit als in anderen Nationen. Dennoch sind viele deutsche Firmen noch sehr unsicher und können sich nicht erklären, warum sie so viel Erfolg in China haben.

In China, »made in Germany« has an incredible resonance. The Chinese have a great longing for authenticity, and find this in German products. German companies think more about the user and user-friendliness than others. Nonetheless, many German companies are uncertain, and cannot explain why they are so successful in China.

»Made in Germany« hat eine unglaubliche Strahlkraft in China mit 1.350 Millionen Menschen, die konsumieren möchten. China war bis in die 80er Jahre keine Konsumgesellschaft, es gab Nutz- aber keine Konsumprodukte. Nun haben all diese Produkte den chinesischen Markt überflutet und die Chinesen haben die Qual der Wahl. Und bei westlichen Firmen gibt es Eldoradofantasien über neue Wachstumsmärkte.

Mittlerweile haben die Chinesen einige Erfahrungen gemacht und sind zum Beispiel von amerikanischen Produkten enttäuscht, weil sie sehr teuer sind und nicht halten, was sie versprechen.

Ein Gegenbeispiel ist Miele. Das Design ist eher konservativ, aber dieses Unternehmen würde nie etwas auf den Markt bringen, was nicht wirklich funktioniert. Gerade das gefällt den Chinesen, weil sie wissen, die Deutschen machen uns nichts vor, die leisten gute Arbeit. Dafür bezahlen sie auch einen Premiumpreis. Sie haben eine große Sehnsucht nach Authentizität, was sie in deutschen Produkten finden. Dort denkt man eher an den Nutzer und die Bedienbarkeit als woanders.

Viele Firmen sind noch sehr unsicher, haben riesige Erfolge, wissen aber nicht, warum. Da sind wir Kulturanthropologen, Soziologen und Zukunftsforscher gefragt, alle Phänomene – Markt, Verbraucher, Konkurrenz, Regularien, Urbanisierung, politische

Linya Int'l Strategic Foresight,
Shanghai

In China, with its 1,350 million people who want to consume, »made in Germany« has an incredible resonance. Until well into the 1980s, China was not a consumer society. There were products for everyday use, but no consumer products. Now all these products have engulfed the Chinese, who now are spoilt for choice. And Western companies are indulging in fantasies about an Eldorado of new growth markets.

In the meanwhile, the Chinese have seen for themselves. For example, they are disappointed with American products, because they are very expensive and do not live up to their promise.

At the other end of the scale, we have Miele. Its design is rather conservative, but Miele would never launch a product that does not work properly. And it is precisely this that the Chinese like. They know that the Germans will not try to pull the wool over their eyes, and that the work they do is good. And they will pay a premium price for this. They have a great longing for authenticity, and find this in German products. German companies think more about the user and user-friendliness than others.

Lebenslauf
Curriculum vitae

Nach seinem Studium in Deutschland, Russland, China und Namibia begann sich der Diplom-Ökonom Christoph Jia mit weltweiten Designmarken, soziologischen Belangen und Konsumwelten in China zu beschäftigen. Er baute die China-Kompetenz für Zukunfts- und gesellschaftliche Marktumfeldforschung (Trendforschung, Szenariotechnik) bei der DaimlerChrysler AG in Berlin auf, deren China Forschungskorrespondent er seitdem ist. 2006 ging er nach Shanghai, wo er sein Unternehmen Linya International gründete. Die strategische Unternehmensberatung berät europäische Konzerne, KMU-Kunden und öffentliche Institutionen zu Land, Markenaufbau und -positionierung. Schwerpunkt ist die Erarbeitung von asiatischen Trends, globalen Zukunftsszenarien und Bereitstellung strategischer Frühaufklärung zu ökonomischen und politischen Themen. Lehrtätigkeit in Deutschland und China.

Following studies in Germany, Russia, China and Namibia, the economics graduate Christoph Jia began to study global design brands, sociological issues and consumer worlds in China. He set up a competence centre for China at DaimlerChrysler AG in Berlin, whose job was to research future and societal market environments (trend research, scenarios). Since then, he has been the centre's China research correspondent. In 2006, he went to Shanghai, where he established his company Linya International. This strategic consultancy advises European companies, SMEs and public institutions about the country, and on how to establish and position brands. Its focus is on identifying Asian trends and global future scenarios, as well as on pinpointing economic and political issues at an early stage. Christoph Jia teaches in Germany and China.

und Umweltsituation – ganzheitlich anzuschauen, um den Firmen ein besseres Verständnis zu vermitteln.

In China sind die Kunden Millionäre oder Milliardäre. Sie haben (noch) keine Markenloyalität, weil sie alles ausprobieren und miteinander kombinieren. Andererseits haben sie eine reiche Farben- und Formensprache und eine jahrtausendealte Tradition, der sie sich aber nicht wirklich zugewandt haben. Diese Tradition wird jetzt wieder entdeckt und mit dem, was von außen kommt, kombiniert. Dies wird eine chinesische Designzukunft werden. Vielleicht spricht man dann von Sinofication, wenn China seine Wurzeln ausbreitet über die Welt.

Wir sind ein kleines Land, aber wir – auch wir Juroren – haben eine große Wirkung auf die Welt und deren Konsummuster. Sehr wichtig ist, nachhaltige Produkte mit einem überzeugenden, klaren Design zu gestalten und sparsam mit Materialien umzugehen. Wir haben eine große Verantwortung und dafür ist der Designpreis sehr wichtig. Wir dürfen nicht nur an Nutzer in Deutschland denken. Das gleiche Produkt wird auch von zahlreichen anderen benutzt oder nachgeahmt über Jahre hinaus, die keine deutschen Konsum- und Lebensgewohnheiten haben. Das müssen deutsche Designer berücksichtigen.

Many companies are still very uncertain. They are extremely successful, but don't know why. This is where we cultural anthropologists, sociologists and futurologists come in. We have to consider every phenomenon holistically – market, consumers, competition, rules and regulations, urbanization and the political and environmental situation – in order to help companies understand better what is going on.

In China, customers are millionaires and billionaires. They have not (yet) developed any brand loyalty, because they are trying everything out, combining different things. On the other hand, they have a rich language of colour and form, as well as a thousand-year tradition that they have never really embraced. This is now returning, and is being combined with the things coming from outside. The future of design will be Chinese. It may be that we will then speak of sinofication, when China spreads its roots across the globe.

We in Germany are a small country, but we – also as judges – have a considerable effect on the world and its patterns of consumption. It is very important to make sustainable products whose design is convincing and clear, as well as to be economical with materials. We have a great responsibility, and this design award has a crucial role to play here. We must not think only of users in Germany. The same product will be used by many other people, or copied, for years to come. These people will not have German consumer habits and lifestyles. German designers have to take account of this.

Prof. Karin Schmidt-Ruhland
Burg Giebichenstein,
Kunsthochschule Halle

Burg Giebichenstein,
Kunsthochschule Halle

Universal Design heißt nicht, ein Produkt speziell für ältere Menschen oder solche mit Beeinträchtigung zu entwickeln, sondern Problemlösungen zu bieten, die einem größeren Kreis nutzen. Es sollte aber so gestaltet sein, dass man ihm die Problemlösung nicht mehr ansieht, dann ist es auch attraktiver für viele: »Design for All«. Es gibt immer mehr Designer, die daran denken.

Universal design is not about developing a product specifically for the elderly or people with disabilities, but about offering solutions that benefit a wider circle of people. Instead, a product should be designed in such a way that the problem it is solving is hidden. This makes it more attractive to use for a large number of people: »design for all«. There are more and more designers who consider this.

Sehr gelungen fand ich den Cafeteria-Stuhl, der stapelbar ist, leicht und modern aussieht. Seine einladende Geste hat mir gefallen. Die in einer gut proportionierten Höhe angebrachten Armlehnen helfen beim Aufstehen besonders denen, die damit Probleme haben. Das sieht man dem Stuhl aber nicht an. Er wirkt formal ganz selbstverständlich und erst in der Benutzung wird der besondere Komfort deutlich. Das ist in bestem Sinne Universal Design und damit auszeichnenswert.

Ich schaue erstens, ob das Produkt selbsterklärend ist und ob es seine Funktion erfüllt, zweitens auf die Form, ob sie einladend ist, mich zum Gebrauch auffordert und mir gefällt. Beides zusammen ergibt ein nützliches und interessantes Produkt. Übertreibungen bei der Gestaltung, um die Funktion zu verdeutlichen, sollte man vermeiden, wenn sie die Beeinträchtigungen ihrer Nutzer hervorheben. Ein Produkt sollte im Gegenteil so gestaltet sein, dass man ihm die Problemlösung nicht mehr ansieht. Dann ist es auch attraktiver für viele in der Nutzung, Stichwort »Design for All«.

Spannend finde ich auch die Dusche OpenSpace von Duravit, natürlich bodengleich, sie ist Universal Design. Es sind einfach Flügeltüren, die das Wasser abhalten. Wenn ich sie einklappe, habe ich zusätzlichen Raum, die Lösung für kleine Bäder. Hinzu kommt, dass eine der Türen beidseitig als Spiegel ge-

I thought the cafeteria chair was very well done. It is stackable and light, and looks modern. I liked its inviting air. The armrests, attached at a well-proportioned height, are especially useful for people who need help when standing up. But the chair doesn't give that impression. Formally, it seems completely matter of fact. It is only when it is used that its special comfort becomes apparent. This is universal design in its best sense, and this makes it prize-worthy.

The first thing I look for is whether the product in self-explanatory and fulfils its function. Second, I look at its form, to see whether it is inviting and asks me to use it, and whether I like it. The two things together result in a beneficial, interesting product. Exaggerated design in order to illustrate function should be avoided if this draws attention to the users' disabilities. Instead, a product should be designed in such a way that the problem it is solv-

Karin Schmidt-Ruhland studierte zunächst Betriebswirtschaft und Soziologie, anschließend Industrial Design an der Hochschule der Künste HdK Berlin (spätere UdK Berlin). Parallel dazu führte sie ihre eigene Agentur für Gestaltung. 1991 übernahm sie eine Lehrtätigkeit an der Universität der Künste UdK Berlin und war dort künstlerische Mitarbeiterin. Von 1998 bis 2003 war sie als wissenschaftliche Mitarbeiterin im Forschungsprojekt »sentha« an der UDK Berlin tätig, wo sie von 2003 bis 2004 auch eine Gastprofessur innehatte. Das legte den Grundstein für ihren Schwerpunkt »Universal Design«. So übernahm sie 2007 die Projektleitung für »Universal Design« beim Internationalen Design Zentrum Berlin und zählt heute zu den wichtigsten Experten in diesem Bereich. Seit 2007 lehrt sie als Professorin für Spiel- und Lerndesign an der Burg Giebichenstein und seit 2010 ist sie Prorektorin für Lehre.

Karin Schmidt-Ruhland studied business administration and sociology before studying industrial design at the Academy of Arts in Berlin. During her studies, she also ran her own design agency. In 1991, she became a lecturer at the Academy of Arts in Berlin, and was an associated artist. From 1998 to 2003 she was a member of the »sentha« research project at the Academy of Arts in Berlin, and was a guest professor at the same institution from 2003 to 2004. This was the foundation for the focal point of her work: universal design. In 2007, she became project manager for universal design at the International Design Centre Berlin. Today, she is regarded as one of the leading experts in this field. She has been a professor at Burg Giebichenstein since 2007, teaching game and learning design. Since 2010, she has been the prorector in charge of teaching there.

nutzt werden kann, sowohl beim Duschen als auch im eingeklappten Zustand, wodurch der Raum sich nochmals vergrößert, was mir sehr gut gefallen hat.

Ich würde mich freuen, wenn noch mehr Produkte eingereicht werden, die den Fokus auf »Mehrgenerationen« haben. Universal Design heißt für mich nicht, ein Produkt speziell für ältere Menschen oder solche mit spezifischer Beeinträchtigung zu entwickeln, sondern Problemlösungen zu bieten, die einem größeren Kreis nutzen. Es gibt immer mehr Designer, die daran denken. Ich würde mir aber wünschen, dass es einfach selbstverständlich ist. Es gibt noch viele Produkte, wo zu wenig an die wirklichen Benutzer gedacht wird.

ing is hidden. This makes it more attractive to use for a large number of people, as befits the ideal of »design for all«.

I also find Duravit's OpenSpace shower exciting. Of course it is flush with the floor: it is universal design. These are simply folding doors that keep water back. When I fold them back, I have extra space – a solution for small bathrooms. What's more, one of the doors can be used on both sides as a mirror, both when showering and when folded away. This enlarges the room even more. I liked that very much.

I would be pleased if even more products were submitted that focus on multi-generational use. For me, universal design is not about developing a product specifically for the elderly or people with disabilities, but about offering solutions that benefit a wider circle of people. There are more and more designers who consider this. However, my wish would be that this was a matter of course. There are still many products where too little thought has gone into the real users.

FOCUS OPEN 2011

Die Gewinner des Wettbewerbs – 11 mal »Focus in Gold« und 80 mal »Focus in Silber«

The winners of the competition – 11 times »Focus in Gold« and 80 times »Focus in Silber«

Hersteller
Manufacturer
Viessmann Werke GmbH & Co. KG
D-35108 Allendorf

Design
Designer
Phoenix Design GmbH & Co. KG
D-70376 Stuttgart

Vertrieb
Distributor
Viessmann Werke GmbH & Co. KG
D-35108 Allendorf

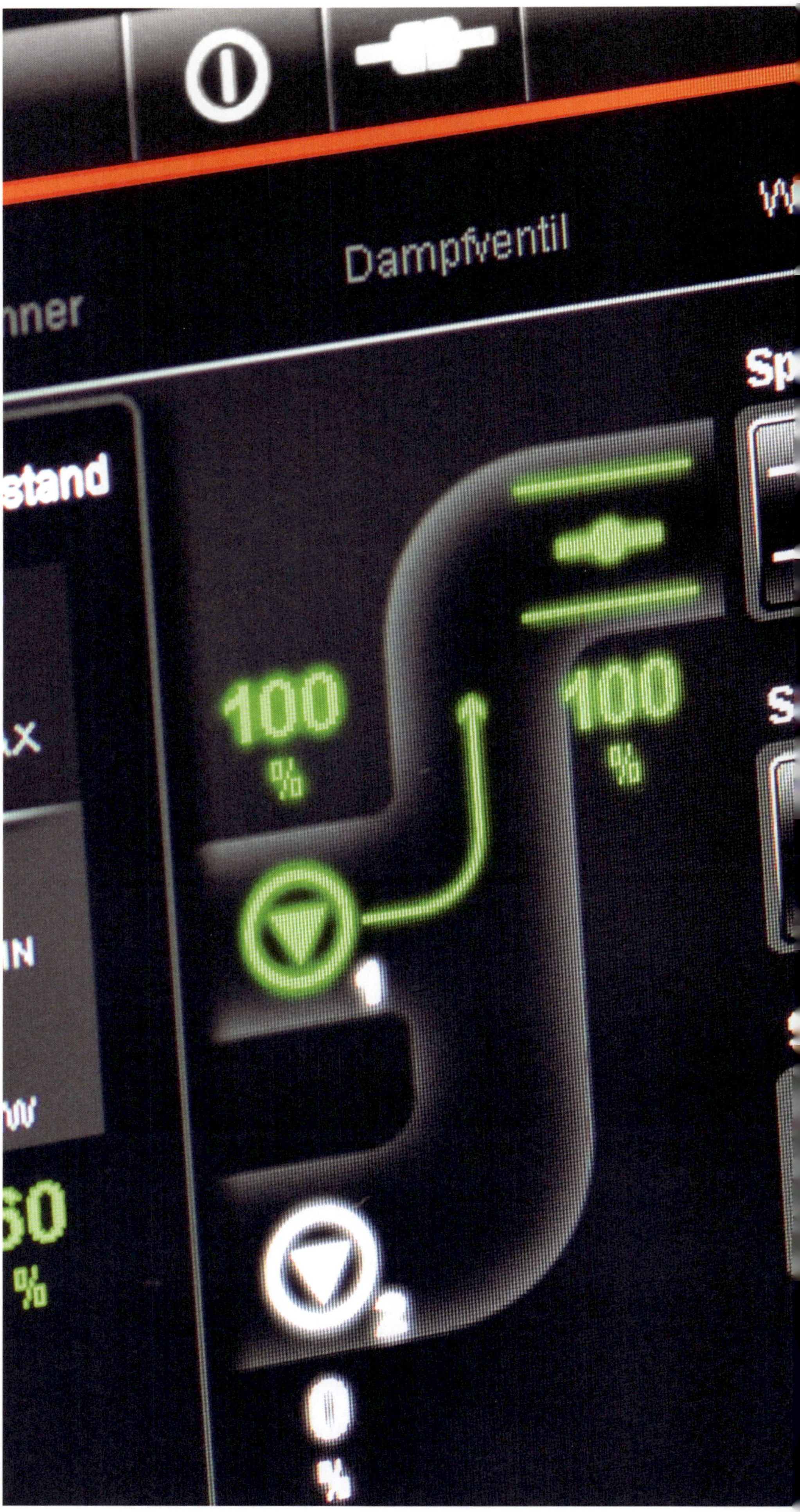

VITO-
CONTROL
VIESS-
MANN

Vitocontrol
Steuer- und Schaltanlage

Control system

Produktion, Montage, Logistik
Gold

Production, assembly, logistics
Gold

022
023

Vitocontrol
Steuer- und Schaltanlage

Control system

Produktion, Montage, Logistik
Gold

Production, assembly, logistics
Gold

Jury
Judges panel

Ein komplexes Profi-Tool, das selbsterklärend und extrem leicht zu bedienen ist: Einfache Icons mit mehreren Funktionen können ausgewählt werden und bieten eine gute Kontrolle. Das zeigt, dass Viessmann neben dem neuen Produktdesign auch im Bereich User-Interface neue Maßstäbe setzt und seine Marktführerschaft unterstreicht.

A complex professional tool that is self-explanatory and extremely easy to use. Simple icons with several functions can be selected, and offer good control. It is not only in its new product design that Viessmann sets new standards and underscores its market leadership, but also in the user-interface area.

Hersteller
Manufacturer
Viessmann Werke GmbH & Co. KG
D-35108 Allendorf

Design
Designer
Phoenix Design GmbH & Co. KG
D-70376 Stuttgart

Vertrieb
Distributor
Viessmann Werke GmbH & Co. KG
D-35108 Allendorf

Vitocontrol setzt mit seinem grafischen Displaylayout im industriellen Maschinenbau-Umfeld neue Maßstäbe. Die Nahtstelle zwischen Mensch und Maschine ist so ausgelegt, dass sie von verschiedenen Nutzergruppen einfach, sicher und intuitiv bedient werden kann. So wurde die Informationsmenge entsprechend für Beobachter, Kesselwarte und Techniker logisch angeordnet, auf ein sinnvolles Maß reduziert und grafisch eindeutig visualisiert.

Über einfache Symbole wie grüne Häkchen verschafft sich der Beobachter einen Überblick über die wichtigsten Funktionen. Kesselwarte und Techniker dagegen finden darüber hinaus auf tiefer liegenden Ebenen alle für sie notwendigen Funktionen.

Grafisch eindeutige Icons sorgen für die Bediensicherheit: So werden die Einstellwerte nicht nur in Zahlen, sondern auch anhand von Balken, Diagrammen und Skalen dargestellt. Dadurch erkennt der Nutzer auf den ersten Blick, ob ein Wert im grünen Bereich ist – oder ob Handlungsbedarf besteht.

Dieses benutzerfreundliche zehn Zoll große Touch Panel wurde für den Vitomax-Industrie-Dampfkessel entwickelt. Es trägt die für Viessmann typischen Corporate-Design-Züge und vermittelt über seine hochwertige Gestaltung, wie beispielsweise das perfekte Finish des Rahmens, die Qualität des Produktes.

Vitocontrol's graphic display layout sets new standards in industrial mechanical engineering. The human-machine interface is designed to allow it to be operated simply, safely and intuitively by various groups of users. Depending on the needs of the observer, janitor and engineer, information flows are logically arranged, reduced to a sensible minimum and visualized in a graphically clear way.

Simple symbols such as green ticks give the observer an overview of the most important functions. The janitor and engineer, by contrast, can access all the more detailed information they need on subsequent levels.

Graphically clear icons ensure operating safety. Setting values are shown not only in numeric form, but also in the form of bars, diagrams and scales. In this way, the user can see at a glance whether a value is within the normal range, or whether there is a need to take corrective action.

This user-friendly ten-inch touch panel was developed for the Vitomax industrial steam boiler. It has the typical characteristics of Viessmann corporate design. Its elegant features, such as the perfect finish of the frame, emphasize the quality of the product.

sserstand
Wasserqualität
im Kessel
isewasserventil
STOP
eisewasserpumpe 1
eisewasserpumpe 2

MENU
Druck
9,7

Vetron
Industrielle Näh- und Schweiß-maschinenbaureihe

Series of industrial sewing and welding machines

Jury
Judges panel

Produktion, Montage, Logistik
Silber

Production, assembly, logistics
Silver

Sehr überzeugende, hochdynamische Grundstruktur, die optisch auf den Hauptaktionspunkt ausgerichtet ist. Alle Bedienelemente befinden sich ebenfalls in diesem Bereich. Zeitgemäße Strukturen, hohe Dynamik. Sehr schön das eigenständige signifikante Erscheinungsbild und das modulare Prinzip.

Very impressive, highly dynamic basic structure, with a visual focus on the main centre of action. All the operating elements are also in this area. Contemporary structures, very dynamic. Its distinctive appearance and modular construction principle are very attractive.

Hersteller
Manufacturer
Xi'anTypical Europe GmbH
D-67657 Kaiserslautern

Design
Designer
Werksdesign
In-house design
Holger Labes,
Bettina Wassermann

Vertrieb
Distributor
Xi'anTypical Europe GmbH
D-67657 Kaiserslautern

Der größte chinesische Hersteller von Industrienähmaschinen gründete eine deutsche Forschungs- und Entwicklungsabteilung, um sich damit das Premium-Segment zu erschließen. So vereint diese Nähmaschinenserie deutsches Hightech-Wissen mit den geringen chinesischen Herstellungskosten.

Die Neuentwicklung verlässt das klassische Prinzip des Nähmaschinenbaus. Statt sämtliche mechanische Bauteile direkt in ein Gehäuse einzubauen, wird der sogenannte »Frame« auf seine tragende Funktion reduziert und die Bauteile in »Technology Modules« zusammengefasst, die jederzeit ausgetauscht werden können.

Durch diese Innovation ist der Kunde in der Lage, die Maschine jederzeit an zukünftige Anforderungen anzupassen. Beispielsweise ist es in wenigen Minuten möglich, eine Flachbett- in eine Langarmmaschine umzubauen. Muss die Nähmaschine repariert werden, kann der Mechaniker das Modul einfach austauschen, um Stillstandszeiten und mögliche Engpässe in der Produktion zu vermeiden.

Vetron zeichnet sich auch durch ihre revolutionäre computergesteuerte Nähdynamik aus. Alle Maschinenparameter sind programmier- und speicherbar. Der Nutzer kann durch das Abrufen der entsprechenden Konfiguration die Maschine einrichten, auch ohne nähtechnische Kenntnisse. Anhand einer kontinuierlichen Materialdickenmessung sowie der momentanen Antriebsdrehzahl werden alle Prozessparameter ermittelt und auf die aktuelle Situation angepasst, wie Stichlänge, Fußdruck, Fadenspannung und Obertransporthub.

The largest Chinese manufacturer of industrial sewing machines has set up a German research and development department and decided to use this as a basis for moving into the premium segment. This series of sewing machines combines German high-tech know-how with low Chinese manufacturing costs.

This new development leaves behind it the classic principle of sewing machine construction. Instead of installing all the mechanical components directly in a housing, the »frame« has been reduced to a supporting function and the components grouped into »technology modules« that can be exchanged at any time.

As a result of this innovation, customers can adjust the machine to any future requirements. For example, within just a few minutes, it is possible to transform a flatbed into a cylinder-bed machine. If the sewing machine has to be repaired, the technician can simply replace a module, and in this way avoid down times and possible bottlenecks in production.

Vetron is also remarkable for its revolutionary, computer-controlled sewing speeds: all machine parameters can be programmed and memorized. By calling up the configuration they need, users can set up the machine without any knowledge of sewing technology. Because material thickness and current drive speed are constantly monitored, all process parameters are recorded and adjusted to the present situation, such as stitch length, knee pressure, thread tension and top feed stroke.

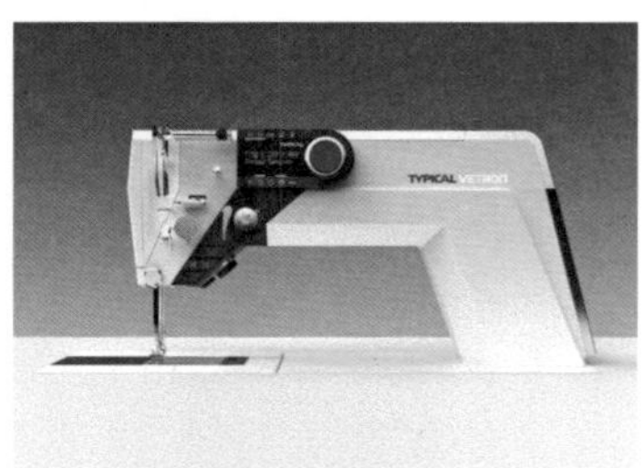

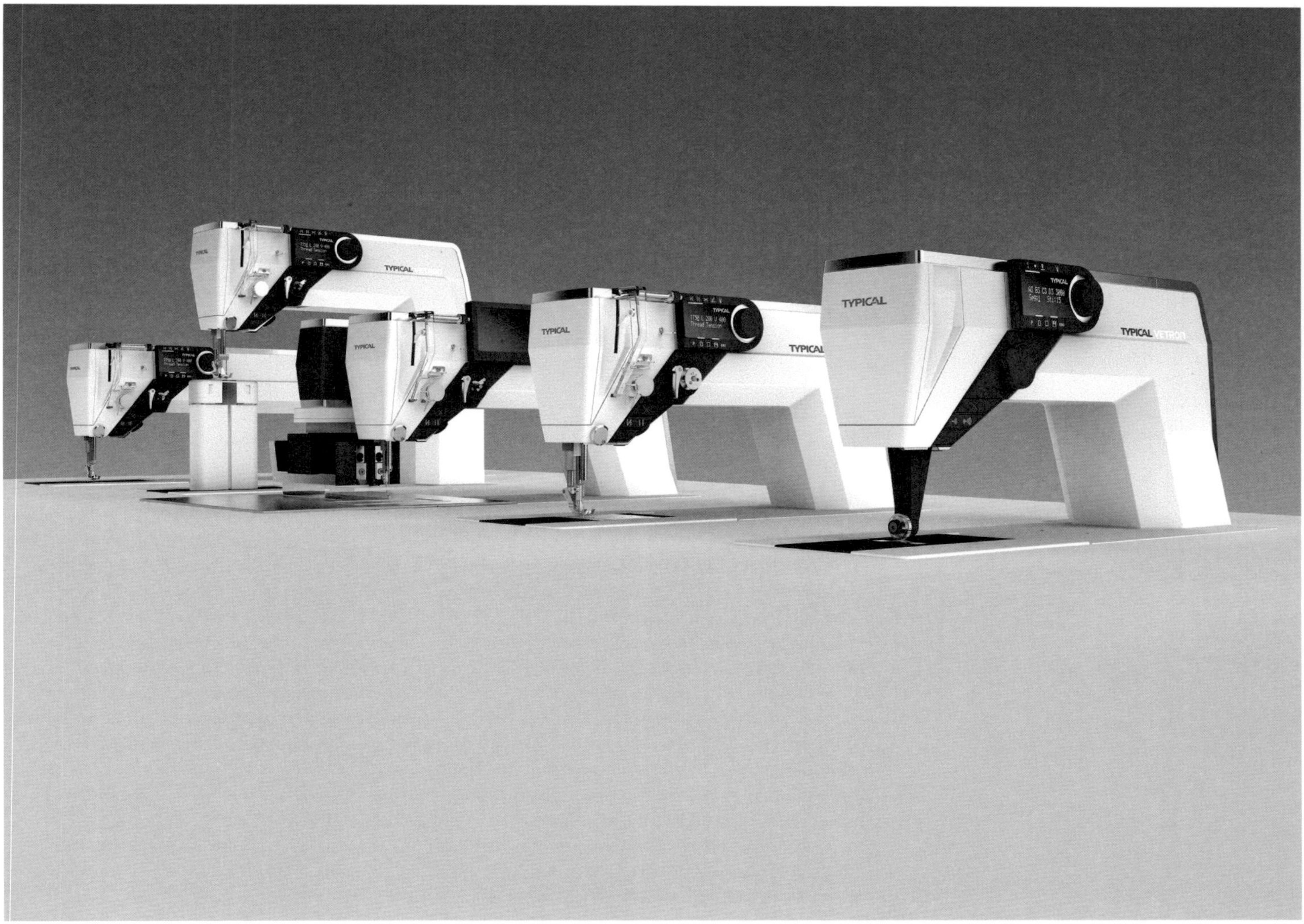
TYPICAL
TYPICAL
TYPICAL
TYPICAL

G-Quick
Schnellmontagesystem für Theatervorhänge

Rapid hanging system for stage curtains

Jury
Judges panel

Produktion, Montage, Logistik
Silber

Production, assembly, logistics
Silver

Sinnvolle funktionale Neuentwicklung, die eine Sisyphos-Arbeit enorm erleichtert und viel Zeitersparnis den Menschen bringt, die die Vorhänge aufhängen. Uns überzeugt zusätzlich, dass dieses Vorhangbefestigungssystem mit dem alten traditionellen System der Ösen kompatibel ist.

Functional new development that makes sense, making a toilsome task much easier and saving the people who hang curtains a lot of time. A further plus point for us is that this curtain hanging system is compatible with the old traditional system of eyelets.

Hersteller
Manufacturer
Gerriets GmbH
D-79224 Umkirch

Design
Designer
Spek Design
D-70565 Stuttgart

Vertrieb
Distributor
Gerriets GmbH
D-79224 Umkirch

Bisher werden im Theater die Vorhänge an Bändern befestigt. Pro Tag wird der Behang durchschnittlich zweimal gewechselt. Das bedeutet, dass täglich bis zu 3.400 Bänder gebunden und ebenso viele Schlaufen gelöst werden müssen – eine enorm zeitaufwändige Tätigkeit.

Dafür wurde nun eine grundlegend neue und zeitgemäße Befestigungsweise von Theatervorhängen entwickelt: das Schnellmontagesystem G-Quick.

Dieses intelligente und innovative System besteht aus einem einteiligen Kunststoffclip mit Filmscharniertechnik, Gummiseil und Haltekugel. Der Kunststoffclip wird an der Vorhangöse festgemacht. Seine Cliphälften sind unterschiedlich lang, um den Öffnungsprozess zu erleichtern. Wird der Clip geschlossen, rastet ein Haltestift deutlich hörbar ein.

Der Clip wird mit dem Gummiseil an der Laststange befestigt, das einfach und schnell über die Haltekugel gestreift wird. Eine vernähte, griffige Gurtbandschlaufe sorgt dabei für die ergonomische Kraftübertragung beim Spannen und Lösen des Gummiseils. Die Kunststoffhaltekugeln unterscheiden sich in ihrer Funktion durch ihre Farbe: Schwarz ist die Norm, Rot markiert die Vorhangmitte.

Mit diesem neuartigen patentierten Vorhangbefestigungssystem ist nun gegenüber der herkömmlichen Befestigungsart eine bis zu dreimal schnellere Montage und Demontage möglich.

Up to now, stage curtains have been attached to cords. On average, curtains are changed twice every day. This means tying up to 3,400 cords every day, and undoing just as many loops – an extremely time-consuming activity.

To solve this problem, a fundamentally new and contemporary way of hanging theatre curtains has been developed: the G-Quick rapid hanging system.

This smart and innovative system comprises a single-component plastic fastener with a film-hinge mechanism, rubber cord and stopper. The plastic fastener is attached to the eyelets on the curtain. The two halves of the fastener differ in length, which makes it easier to open. When the fastener is closed, a locking pin clicks audibly into place.

The fastener is attached to the curtain track using a rubber cord that is simply and quickly slid over the stopper. A sewn, easy-to-grip webbing loop ensures that force is transferred smoothly when tightening and loosening the rubber cord. The plastic stoppers' colour is a key to their different functions. Black is standard, while red marks the centre of the curtain.

Thanks to this innovative, patented curtain hanging system, hanging and taking down curtains is as much as three times faster than with conventional hanging systems.

Ø 50-70 mm
Ø 50-70 mm

MPS
Magnetsensor

Magnetic cylinder sensor

Jury
Judges panel

Produktion, Montage, Logistik
Silber

Production, assembly, logistics
Silver

Kleine Ursache, große Wirkung: Ein sehr sauber über die Anwendung entwickeltes Produkt, das eine hohe Flexibilität an einem wichtigen Punkt mit einfachsten Mitteln bietet. Man kann es frei einklicken in ein bestehendes Profil und durch einfachste Schraubenelemente befestigen. Eine perfekte Lösung für eine eigentlich relativ schwierige Aufgabe. Und man kann es leicht nachjustieren.

Small cause, great effect. The application has been used as the basis for developing a very neat product, which offers a lot of flexibility at an important point, using the simplest means. It can simply be clicked in to an existing slot and fixed using simple tightening elements. A perfect solution for what is in fact a relatively difficult task. And it is simple to adjust subsequently.

Hersteller
Manufacturer
Sick AG
D-79183 Waldkirch

Design
Designer
Synapsis Design GmbH
D-70190 Stuttgart

Vertrieb
Distributor
Sick Vertriebs-GmbH
D-40549 Düsseldorf

Um Sondermaschinen in der Verpackungsindustrie oder bei sonstigen Anwendungen millimetergenau zu positionieren, war man bisher auf relativ teure elektrische Linearantriebe angewiesen. Mit dem Miniatur-Magnetsensor MPS wird dies nun auch für wesentlich preiswertere hydraulische oder pneumatische Kolben möglich, die bislang nur zwei Schaltzustände – auf oder zu – kannten. Voraussetzung hierfür ist die kurze Reaktionszeit von 0,1 Millisekunden und die exakte magnetische Detektion der Kolbenposition (0,5 mm).

Eine herausragende Innovation des MPS ist seine Drop-in-Fähigkeit, das heißt, der Sensor wird von oben in die Nut eingesetzt und nicht mehr seitlich eingefädelt. Das vereinfacht den Service in komplexen Anlagen enorm.

Die patentierte Diagonalverschraubung mit konischen Gewindestiften gleicht Maßunterschiede der Nut bei unterschiedlichen Kolbenherstellern aus. Sie steckt in Endkappen aus hochfestem Kunststoff, die nur minimale Totzonen beim Messen ergibt. Durch den eingezogenen Kabelauslass kann der Sensor zudem bis auf Anschlag positioniert werden.

Der asymmetrische Querschnitt macht das Einsetzen leichter und minimiert den Platzbedarf in der entsprechenden Einbausituation. Die Formgebung fördert intuitiv richtiges Handling. Bisher gibt es acht Längen für die gewünschten Stellbereiche der Kolben. Über den Teachknopf und die Zustands-LED werden nach der Montage des MPS Anfangs- und Endpunkt eingelernt. Dann steht einer genauen Positionierung nichts mehr im Wege.

For millimetre precision when positioning special-purpose machinery in the packaging industry or other applications, relatively expensive linear-motion drives have so far been the only option. Thanks to the MPS miniature magnetic cylinder sensor, this is now also possible with far more cost-effective hydraulic and pneumatic pistons, which up to now have only had two operating modes – open or closed. The condition for this is a short reaction time of 0.1 milliseconds and precise magnetic detection of piston position (0.5 mm).

One outstanding innovative feature of the MPS is its drop-in mounting. In other words, the sensor is inserted into the groove from above, and no longer threaded in from the side. This makes the servicing of complex machinery considerably easier.

The patented diagonal screw fitting with set screws offsets dimensional differences between the grooves of different piston manufacturers. It is inserted in end caps of high-strength plastic, which means there are only minimal dead zones when measuring. Its retracted cable outlet also means that the sensor can be positioned right against the end-stop.

The sensor's asymmetric section makes mounting easier, and keeps the amount of space needed for installation to a minimum. Its design encourages intuitively correct handling. Up to now, eight lengths are available for the desired piston settings. Using the teach button and status LED, the zero and final point of the MPS can be taught in following assembly. There is then nothing standing in the way of accurate positioning.

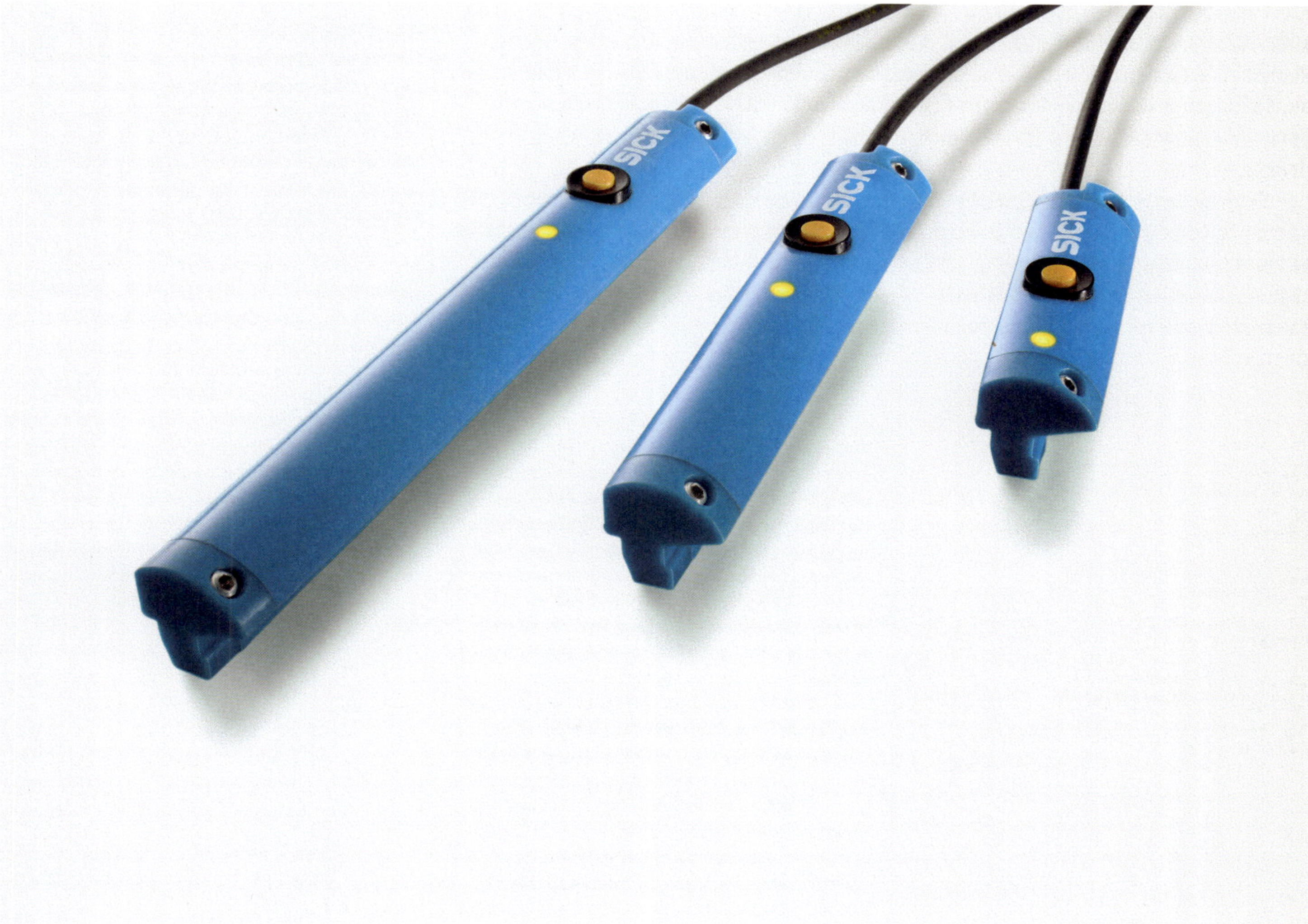

Focus^{3D}
3D-Laserscanner

3D laser scanner

Produktion, Montage, Logistik
Silber

Production, assembly, logistics
Silver

Jury
Judges panel

Ein Produkt, das seine hoch innovative Funktionalität, nämlich das räumliche Scannen von Großflächen, Gebäuden et cetera auch über seine Struktur widerspiegelt. Ganz klare Zentrierung von statischen Elementen auf den Hauptfunktionsbereich. Materialität und Farbwahl sind ebenso gut gelöst wie Branding und Naming. Ein rundum gelungenes Produkt.

A product whose structure reflects its highly innovative function, which is the spatial scanning of large surfaces, buildings, etc. Its static elements have been clearly focused on its main function. Good choice of materials and colour scheme, and branding and naming are also well executed. All in all, a well executed product.

Hersteller
Manufacturer
Faro Scanner Production GmbH
D-70825 Korntal-Münchingen

Design
Designer
Axel Ruhland Produktgestaltung
D-70190 Stuttgart

Vertrieb
Distributor
Faro Scanner Production GmbH
D-70825 Korntal-Münchingen

Klein, leicht und benutzerfreundlich – das sind nur einige der Eigenschaften des Focus^{3D}. Er ist ein Hochgeschwindigkeits-3D-Scanner zur detaillierten Messung und Dokumentation, der sich durch seine einfache Bedienung und seine kompakte Bauweise für die verschiedensten Anwendungen einsetzen lässt: von der dreidimensionalen Gebäudedokumentation und Bauüberwachung über Denkmalschutz und Reverse Engineering bis hin zur forensischen Tatortdokumentation.

Mithilfe von Lasertechnologie erzeugt das Gerät innerhalb weniger Minuten beispiellos detaillierte dreidimensionale Bilder von komplexen Umgebungen und Geometrien.

Durch sein intuitives Bedienkonzept mit Touchscreen-Display ist der Focus^{3D} so einfach zu bedienen wie eine Digitalkamera. Zudem arbeitet das Gerät vollkommen autark, das heißt man benötigt weder Zusatzgeräte noch Kabel oder Laptop.

Dank seiner minimalen Abmessungen und seines geringen Gewichts sowie seiner Touch-Schnittstelle kann man unkompliziert mit dem Focus^{3D} arbeiten und im Vergleich zu herkömmlichen Scannern sogar bis zu 50 Prozent Scanzeit einsparen.

Small, light and user-friendly – these are just a few of the characteristics of Focus^{3D}. It is a high-speed 3D scanner for detailed measurement and documentation. Its simple operation and compact construction mean it can be used for a broad range of applications: from three-dimensional construction documentation and site supervision, to monument conservation and reverse engineering, to forensic documentation at crime scenes.

With the help of laser technology, the appliance generates exceptionally detailed, 3D images of complex surroundings and geometries within a few minutes.

Intuitive to operate using a touchscreen display, Focus^{3D} is as easy to use as a digital camera. In addition, the appliance works completely independently – there is no need for additional appliances, cables or laptops.

Thanks to its very small size and light weight, as well as its touchscreen interface, working with Focus^{3D} is uncomplicated. And compared with conventional scanners, scanning time can be reduced by as much as 50 percent.

FARO
FOCUS3D

PP
Zapfpistole

Filling nozzle

Jury
Judges panel

Produktion, Montage, Logistik
Silber

Production, assembly, logistics
Silver

Ergonomische Überarbeitung eines Standardtools zum Zapfen von Säuren und Laugen. Sehr durchdacht, ergonomisch ausgereift, sauberer Drehpunkt, gut gestaltetes Grifftool. Minimalistisch, leicht, günstig, überzeugend. Und die richtige Farbwahl: bewegliches Element rot, statisches schwarz. Professionell gemacht.

Ergonomic rethink of a standard tool for filling acids and alkalis. Very well conceived, ergonomically mature, perfect fulcrum, well designed for gripping. Minimalistic, light, cost-effective, convincing. And the right choice of colour: the movable element is red, the static part black. Professionally done.

Hersteller
Manufacturer
Flux-Geräte GmbH
D-75433 Maulbronn

Design
Designer
Teams Design GmbH
D-73728 Esslingen

Vertrieb
Distributor
Flux-Geräte GmbH
D-75433 Maulbronn

Die Zapfpistole, die für eine schnelle und sichere Förderung aller Säuren und Laugen eingesetzt wird, wurde ergonomisch optimiert, und sie zeichnet sich durch eine besonders langlebige Konstruktion aus.

So benötigt man dank des ergonomisch geformten Handhebels weniger Kraft. Ein Sicherheitsbügel schützt davor, dass der Handhebel aus Versehen betätigt wird. Das Zubehör für die Auslaufseite – verschiedene Auslauftüllen für unterschiedliche Einsätze – ist für alle Anwendungen leicht adaptierbar.

Beim Ventil wurde die Abdichtung optimiert. Das hat zur Folge, dass nicht nur die Bedienung einfacher und die Betriebssicherheit verbessert, sondern auch der Durchfluss erhöht und die Dosierbarkeit feiner und genauer wurde.

Außerdem besitzt die Zapfpistole ein neuartiges Drehgelenk mit umlaufenden Kugeln. Dies garantiert auch unter Druck absolute Dichtheit und sorgt für eine sichere Handhabung.

Die Zapfpistole, die aus Polypropylen besteht, lässt sich auch einfach und leicht dank der großen Aufhängeöse aufbewahren. Durch ihre Farbwahl und Formgebung integriert sie sich harmonisch in das Flux-Erscheinungsbild.

This filling nozzle, which is used for the fast and secure delivery of acids and alkalis of all kinds, has been ergonomically optimized. Its striking feature is the way it has been designed for an especially long life.

The ergonomically shaped trigger means that less strength is needed. A safety catch prevents the trigger from being pulled accidentally. The discharge accessories – various discharge spouts for a broad range of uses – can easily be adapted to applications of all kinds.

The valve seal has been optimized. As a result, not only is operation simpler and operating safety higher, but flow is also increased and metering is more accurate.

In addition, the filling nozzle features an innovative pivot joint with rotating balls. This ensures an absolutely tight seal even under pressure, and thus safe handling.

Made of polypropylene, the filling nozzle can be stowed away simply and easily thanks to its large suspension eyelet. Thanks to its colour scheme and design, it harmonizes well with Flux corporate design.

FLUX

IntelliTop® 2.0
Steuereinheit für Prozessventile

Control head for process valves

Jury
Judges panel

Produktion, Montage, Logistik
Silber

Production, assembly, logistics
Silver

Das konventionelle Problem war die räumliche Trennung von Signallampe und Ventil. Hier dagegen wurde die Farbsignalleuchte in das System integriert. Das bietet dem Nutzer eine eindeutige Botschaft: grün in Ordnung, gelb Warnung und rot Störung. Eine ganz klare Sache, die zur sicheren Bedienung komplexer Anlagen beitragen kann. Hinzu kommt die gelungene Gestaltung.

The problem with conventional solutions was that the signal lamp and valve were spatially separated. Here, by contrast, the colour signal display has been integrated into the system. This gives the user a clear message: green for OK, yellow for warning and red for malfunction. A crystal-clear display that can help make the operation of complex machinery safe. And it's elegantly designed as well.

Hersteller
Manufacturer
Südmo Components GmbH
D-73469 Riesbürg

Design
Designer
Modus Product Design GmbH
Alexander Benn, Harry Eppinger
D-70435 Stuttgart

Vertrieb
Distributor
Südmo Components GmbH
D-73469 Riesbürg

Bei Produktionsanlagen in der Getränke-, Lebensmittel- und Pharmaindustrie ist es von großer Bedeutung, für welche Automatisierungstechnik der pneumatischen Anlagenbauteile man sich entscheidet. Bisher wurden bei einer zentralen Steuerung die Prozessventile von einem separaten Schaltschrank mit Magnetventilinseln angesteuert und über zusätzliche Rückmelder überwacht.

Ganz anders dagegen bei der Steuereinheit IntelliTop® 2.0. Sie vereint in einem einzigen Gerät die Ansteuerung sowie die Überwachung von Prozessventilen. Durch die dezentral auf den Prozessventilen angeordneten Steuerköpfe können Schlauch- und Kabellängen reduziert und der Anlagenaufbau übersichtlicher werden.

Eine Rundum-Stellungsanzeige visualisiert dem Bediener schon von weitem den Schaltzustand des Prozessventils sowie auftretende Fehler. Die Signalfarben Grün, Gelb und Rot sind hierbei kundenspezifisch wählbar.

Das Herz der IntelliTop® 2.0 ist das Wegmesssystem zu Erfassung von bis zu drei Schaltstellungen des Prozessventils. Eine einfache Programmierung mittels dreier Teach-in-Tasten garantiert eine schnelle und prozesssichere Inbetriebnahme. Außerdem können die Prozessventile bei geschlossenem und somit geschütztem Steuerkopf gewartet werden.

In the production lines in the beverages, foodstuffs and pharmaceuticals industries, the decision as to what automation technology to use for the pneumatic components is extremely important. In the central control solution used up to now, the process valves have been controlled from a separate switch panel using magnetic valve terminals, and monitored via additional feedback modules.

With the IntelliTop® 2.0, this is completely different. In a single appliance, it combines control and monitoring of process valves. By having locally installed control heads mounted on the process valves, hoses and cables can be shortened, and system design simplified.

Valve position and faults can be seen from any angle and distance thanks to an all-round display. Customers can assign the green, yellow and red signals as desired.

The heart of IntelliTop® 2.0 is its path measuring system, which allows up to three process-valve switching positions to be tracked. Three teach-in buttons make programming easy, and ensure fast and reliable process start-up. In addition, the process valves can be serviced when the control head is closed, and thus protected.

Pentair
Pentair
Intelli Top 2.0
Muster 16
598425

ErgoStar
Stiftschlüsselsatz mit Halter

L key set

Jury
Judges panel

Produktion, Montage, Logistik
Silber

Production, assembly, logistics
Silver

Ein Produkt, das ganz klar widerspiegelt, dass sich der Entwickler intensiv mit der Nutzung beschäftigt hat. Das automatische Herausdrehen beim Bewegen von einem Stiftschlüssel ermöglicht gleichzeitig die Entnahme verschiedener Schlüssel. Dieser Stiftschlüsselhalter ist wunderbar aus der Funktion heraus entwickelt worden.

A product that clearly shows that its developer thought hard about its use. The automatic turning movement when one key is moved allows various keys to be extracted at the same time. The key holder has been wonderfully developed on the basis of its function.

Hersteller
Manufacturer
Wiha Werkzeuge GmbH
D-78136 Schonach

Design
Designer
Werksdesign
In-house design
und
Ahape Design
D-72202 Nagold

Vertrieb
Distributor
Wiha Werkzeuge GmbH
D-78136 Schonach

Genial und einfach zugleich ist die Konstruktion des ErgoStar; denn nur auf den ersten Blick sieht er aus wie ein herkömmlicher Stiftschlüsselhalter.

Doch schon bei der ersten Anwendung merkt man, warum er anders ist: Man muss nicht mehr umständlich die Schlüssel entnehmen, sondern nur einen beliebigen Stiftschlüssel im Halter drehen. Und wie durch eine magische Hand drehen sich alle Schlüssel gleichzeitig zur Seite und können bequem entnommen werden.

Möglich ist dieser Komfort und diese Übersichtlichkeit durch einen patentierten innovativen Öffnungsmechanismus, der dafür sorgt, dass – sozusagen »mit einem Dreh« – mit einer fließenden Bewegung alle Stiftschlüssel zur komfortablen Entnahme bereit sind.

Über die gelaserte, auffällige Angabe der Größe ist der gesuchte Schlüssel schnell zu identifizieren. Eine wirkungsvolle Klemmung sorgt dafür, dass die Schlüssel – sowohl bei geschlossenem als auch geöffnetem Halter – nicht herausfallen können. Ein weiterer Bonus ist das besonders kompakte Design und die handliche Bauweise, wodurch ErgoStar in jede Werkbank und jeden Werkzeugkasten passt.

ErgoStar's design is brilliant and simple at the same time. Only at first glance does it look like a normal L key holder.

Right from the first use, you can see what is different about it: extracting the keys is no longer a fiddly process. All you have to do is turn any of the keys in the holder. As if by magic, all the other keys turn at once to the side, and can easily be extracted.

This convenience and neatness is due to a patented innovative opening mechanism, which ensures that all the keys can be extracted easily at one turn.

With key size engraved clearly by laser, the key that is needed can quickly be identified. An effective clamp ensures that the keys cannot fall out, either when the holder is open or closed. One further benefit is the especially compact design and handy size, which means that ErgoStar fits into every workbench and toolbox.

wiha

Slim-Familie
VDE-Schraubendreher

VDE screwdrivers

Produktion, Montage, Logistik
Silber

Production, assembly, logistics
Silver

Jury
Judges panel

Ein großes Problem, das die herkömmlichen Schraubendreher haben, ist, dass die Verdickung für die Befestigung des Bit vorne ist. Das heißt, man kommt nicht an verborgene Schrauben. Hier wurde einfach das Problem mit einer intelligenten Schnellverbindung nach hinten verlegt. Die Isolierung wurde in die Klingen bis zu den Bits integriert. Haptisch sehr wertig.

One big problem with conventional screwdrivers is that their cross-section thickens at the end where the bit is attached. This means you can't get at every screw. Here, the problem has simply been moved further up the screwdriver, using a smart quick-release connection. They are isolated all the way down to the bits. Very good haptic quality.

Hersteller
Manufacturer
Wiha Werkzeuge GmbH
D-78136 Schonach

Design
Designer
Werksdesign
In-house design

Vertrieb
Distributor
Wiha Werkzeuge GmbH
D-78136 Schonach

Neue Maßstäbe setzt die Slim-Familie dank einer neuen innovativen Technologie – Slim-Technology. So einfach wie genial: Die oft störende Isolierung wurde hier in die durchgehend gehärteten Klingen integriert. Das erlaubt uneingeschränkte Zugänglichkeit zu den Befestigungselementen bei maximaler Sicherheit.

So konnte der Gesamtdurchmesser der Slim-Klingen im kritischen Arbeitsbereich um 11 bis maximal 33 Prozent reduziert werden. Hierdurch sind die Slim-Klingen außergewöhnlich schlank, aber nach wie vor extrem leistungsstark. Sie übertreffen sogar die VDE-Drehmomentanforderungen mehrfach.

Der Vorteil für Elektriker: GS und VDE zertifizierte Sicherheit in hoher Qualität für bislang unzugängliche Verschraubungsstellen.

Alle einzelnen Mitglieder der Slim-Familie sind sicher, kompakt und handlich. Sie bestechen durch hervorragende Haptik und Ergonomie. Die vollständig isolierten VDE-SlimBits können für das VDE-Drehmoment-Programm SlimTorque sowie für das Standard-VDE-Wechselbit-System SlimVario verwendet werden. Der Anwender bekommt damit ein kostensparendes, flexibles Set mit zwei Handgriffen und einem Satz VDE-SlimBits, die je nach Griff für Drehmomentanwendungen oder Standardverschraubungen genutzt werden können.

Damit kann Gewicht und Anzahl der Werkzeugteile reduziert werden – ein großer Vorteil gerade im Wartungsbereich und im mobilen Einsatz.

Thanks to its innovative new slimTechnology, the slim product family sets new standards. As simple as it is clever, this technology integrates the frequently bothersome insulation into the through-hardened bits. This allows unrestricted access to the bits, while preserving maximum safety.

It also means that the total diameter of the slim bits in the critical working area has been reduced by between 11 and 33 percent. Not only are the bits unusually slender, their performance also remains extremely good. They even exceed the torque requirements of the VDE (German electrical standards association) several times over.

For electricians, the benefits are GS and VDE-certified safety, high quality, and the possibility to reach previously inaccessible points.

All the members of the slim family are safe, compact and handy. Their haptic and ergonomic features are excellent. The completely insulated VDE slimBits can be used with the slimTorque VDE torque program and the slimVario standard VDE exchangeable bit system. What users get is a cost-saving, flexible set with two handles and a set of VDE slimBits. Depending on the handle fitted, they can be used for torque applications or standard tightening work.

This allows the weight and number of tools to be reduced – a great advantage for maintenance work and when on the move.

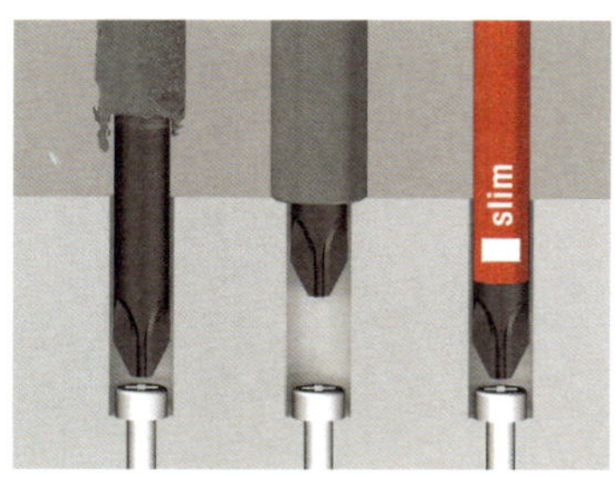

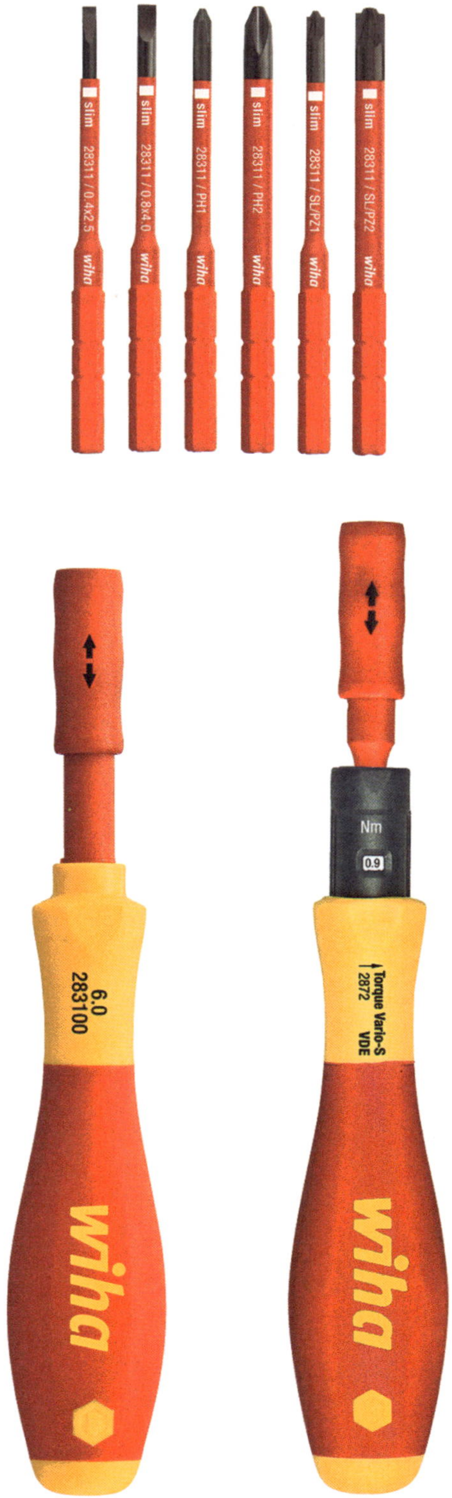
slim
28311 / 0.4x2.5
wiha
slim
28311 / 0.8x4.0
wiha
slim
28311 / PH1
wiha
slim
28311 / PH2
wiha
slim
28311 / SL/PZ1
wiha
slim
28311 / SL/PZ2
wiha
6.0
283100
wiha
Nm
0.9
Torque Vario-S
2872
VDE
wiha

Jury
Judges panel

TP1011
Gerätesatz (Schulungstool)

Equipment set (training tool)

Produktion, Montage, Logistik
Silber

Production, assembly, logistics
Silver

Ein Schulungstool zum Aufbau von elektronischen Schaltkreisen für die Aus- und Weiterbildung. Dieses Produkt spiegelt wider, dass die Designkultur von Festo sich nicht nur nach außen, sondern auch nach innen wendet und eine saubere Gestaltung im Bereich der Ausbildung ihrer Nachwuchskräfte einsetzt. Das ist ein klares Zeichen von Design-Bewusstsein.

An educational tool for assembling electronic circuits in vocational and further training. The product shows how Festo's design culture is not only directed at the outside world but also internally at the company itself. Elegant design is used when training young people. It is a sign of design consciousness.

Hersteller
Manufacturer
Festo Didactic GmbH & Co. KG
D-73770 Denkendorf

Design
Designer
Festo AG & Co. KG
Karoline von Häfen
D-73734 Esslingen

Vertrieb
Distributor
Festo Didactic GmbH & Co. KG
D-73770 Denkendorf

Das TP1011 ist für den Einsatz in der Aus- und Weiterbildung konzipiert. Es ermöglicht den Aufbau von Versuchsanordnungen, um Lehrinhalte in der Elektrotechnik und in der Elektronik sowie in der Digital- und Regelungstechnik zu vermitteln.

Auf einem Steckfeld, das über ein Netzteil mit Energie versorgt wird, können aus einzelnen Steckbauteilen auch komplexe Schaltungen schnell und einfach aufgebaut werden.

Die Steckbauteile sind mit Sicherungstechnik ausgestattet und nach didaktischen Gesichtspunkten gestaltet: So korrespondiert das aufgedruckte Schaltsymbol mit dem tatsächlich verwendeten Bauteil auf der Platine, das durch das transparente Fenster sichtbar ist.

The TP1011 is designed for use in vocational and further training. It allows experiments to be performed that demonstrate principles of electrical engineering and electronics, as well as of digital and closed-loop technology.

By inserting individual plug-in parts into a patch panel that is supplied with power from a mains unit, even complex circuits can be quickly and easily assembled.

The plug-in parts are equipped with safety systems and designed according to educational criteria: the printed switch symbol, for example, corresponds with the actual shape of the component on the printed circuit board, which is visible through a transparent window.

FESTO
5V/200mA
0...25V/0.3A
0V
Output

Hersteller
Manufacturer
Dürr Dental AG
D-74321 Bietigheim-Bissingen

Design
Designer
Formstudio Merklepark GbR
D-70178 Stuttgart

Vertrieb
Distributor
Dürr Dental AG
D-74321 Bietigheim-Bissingen

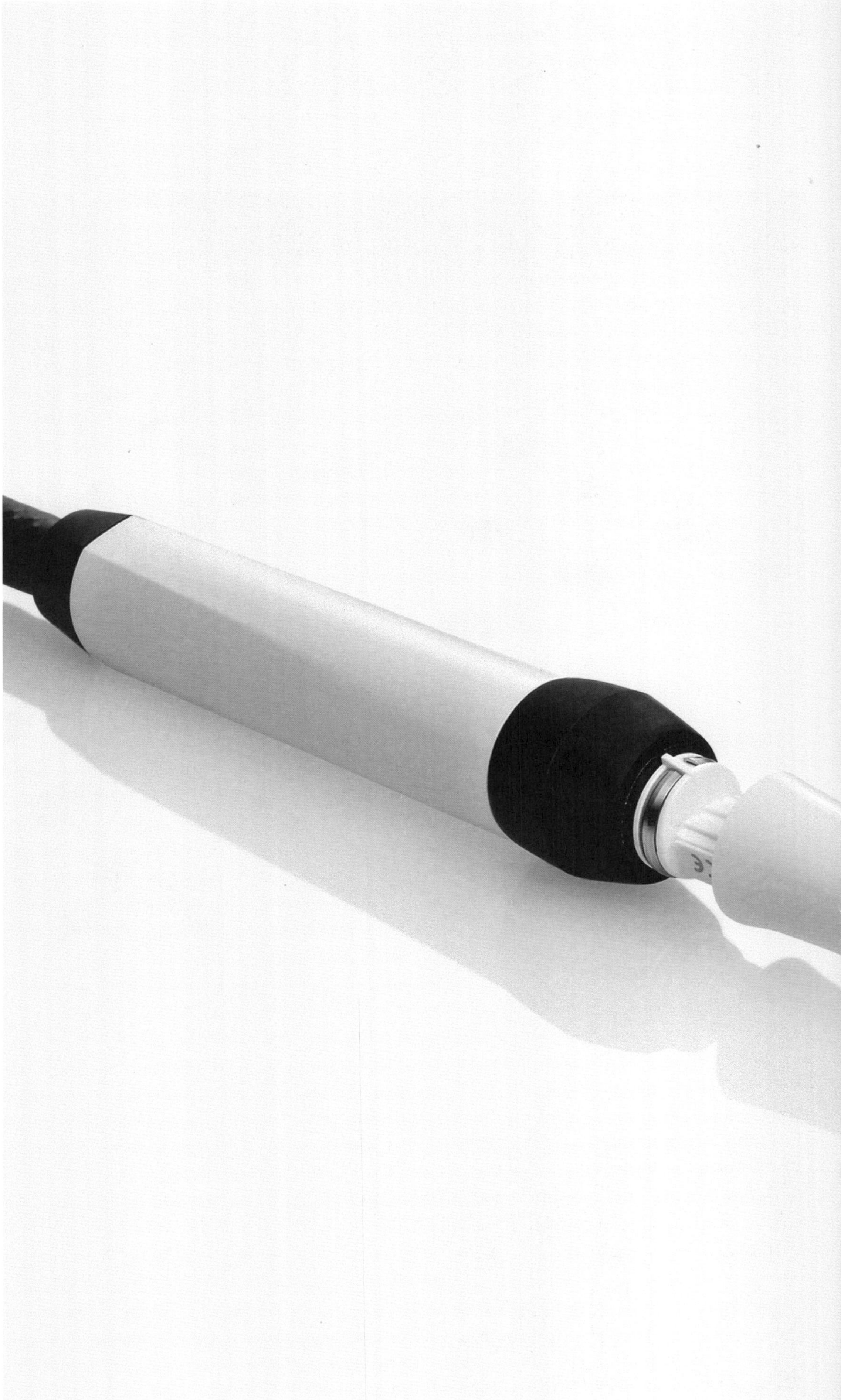

VISTA CAM IX DUERR DENTAL

VistaCam iX
Intraoralkamera

Intraoral camera

Medizin, Rehabilitation, Geriatrie
Gold

Medicine,
rehabilitation, geriatric medicine
Gold

VistaCam iX
Intraoralkamera

Intraoral camera

Medizin,
Rehabilitation, Geriatrie
Gold

Medicine,
rehabilitation, geriatric medicine
Gold

Jury
Judges panel

Ein Spitzenprodukt. Wer hat sich beim Zahnarzt nicht schon gefragt, ob die entsprechende Diagnose auch wirklich stimmt. Anhand dieser Kamera kann man während der Behandlung den Mundraum sehen und die Arbeit des Arztes verfolgen. Diese Kamera ist für alle Nutzergruppen spannend, auch für Kinder, denen man auf diese »bildhafte« Weise die Wichtigkeit der Zahnpflege näher bringen kann. Das beiliegende Kommunikationsmaterial ist ebenfalls gut gestaltet.

An outstanding product. Who hasn't wondered whether what the dentist says is really true? Here, I can see exactly what is happening to my teeth. We think the camera is an exciting product for all user groups – also for children. You can make plaque visible, for example, and show them where caries is beginning to develop and where they haven't cleaned their teeth properly. The accompanying information material has also been well designed.

Hersteller
Manufacturer
Dürr Dental AG
D-74321 Bietigheim-Bissingen

Design
Designer
Formstudio Merklepark GbR
D-70178 Stuttgart

Vertrieb
Distributor
Dürr Dental AG
D-74321 Bietigheim-Bissingen

VistaCam iX bietet Patientenkommunikation, Kariesdiagnostik und Dokumentation in einem. Die Intraoralkamera nimmt Bilder im Mund auf und zeigt sie auf einem Monitor. Damit kann der Zahnarzt Strukturen, wie feine Risse, und krankhafte Veränderungen, die mit bloßem Auge nur schwer zu sehen sind, frühzeitig erkennen. Und der Patient kann besser nachvollziehen, welche Krankheit er hat und welche Behandlungsmaßnahmen notwendig sind.

Der Proof-Wechselkopf macht Karies und Plaque sichtbar. VistaCam iX vereint intraorale Kamerabilder und Fluoreszenzaufnahmen. Die Stoffwechselprodukte kariogener Bakterien werden vom violetten Licht der LEDs angeregt und leuchten rot, gesunder Schmelz hingegen sendet grünes Licht aus. Die Software DBSWIN ermöglicht eine präzise Lokalisation und Auswertung der Kariesaktivität am Bildschirm.

Die Intraoralkamera zeichnet sich nicht nur durch hohe Funktionalität aus, sondern erfüllt auch alle Ergonomieansprüche. Die kreisrunde Anordnung der LEDs sorgt für optimale und homogene Ausleuchtung. Ein Bewegungssensor schaltet die Kamera an und ab.

VistaCam iX kann in nahezu jede Praxisumgebung integriert werden, auch ohne PC mit der Stand-alone Version. Sie lässt sich direkt am Handstück entkoppeln. Dadurch ist sie bestens geeignet für Praxen mit mehreren Behandlungsplätzen.

VistaCam iX combines patient communication, caries diagnosis and documentation. This intraoral camera takes pictures in the patient's mouth and displays them on a monitor. This allows the dentist to recognize at an early stage structures such as fine fractures and pathologic changes, which are very difficult to see with the naked eye. And patients can better understand what condition they have and what treatment is necessary.

The proof head makes caries and plaque visible. VistaCam iX combines intraoral camera images and fluorescent imaging. The metabolic products of the bacteria that cause caries are excited by the LEDs' violet light and send out a red glow, while healthy enamel glows green. DBSWIN software allows the caries activity to be precisely localized and analyzed on-screen.

The intraoral camera is not only highly functional, it also satisfies all the demands of ergonomics. The circular arrangement of the LEDs ensures the best possible, homogeneous illumination. A motion sensor switches the camera on and off.

VistCam iX can be integrated into nearly every surgery environment. Its stand-alone version can also be used without a PC. It can be disconnected directly at the handpiece. This makes it ideal for surgeries with several treatment rooms.

Medical
Medizinische Handzahnbürste

Medical manual toothbrush

Medizin,
Rehabilitation, Geriatrie
Silber

Medicine,
rehabilitation, geriatric medicine
Silver

Jury
Judges panel

Abweichend von herkömmlich geformten Zahnbürsten wird hier Zweikomponenten-Technologie auf eine sehr reduzierte und freundlich anmutende Art und Weise eingesetzt. Puristisches Design. Keine scharfen Kanten. Reinigungsfreundlich. Die Serie umfasst drei verschiedene Zahnbürsten für drei unterschiedliche Anwendungsfelder.

Unlike conventionally shaped toothbrushes, it uses dual-component technology in a very reduced and friendly way. Puristic design. No sharp edges. Easy to clean. The series comprises three different toothbrushes for three different areas of application.

Hersteller
Manufacturer
Interbros GmbH
D-79677 Schönau

Design
Designer
Stotz-Design.com
GmbH & Co. KG
D-42285 Wuppertal

Die Zahnbürsten sind speziell für medizinische Anwendungsbereiche gestaltet. Diesem hohen Anspruch wird das strenge und zurückhaltende Design und die technisch anmutende Formgebung gerecht. Die Zahnbürsten zeichnen sich durch eine glatte, schlanke Linienführung und leicht geschwungene Formen aus.

Glasklares PET und der sparsame Einsatz von transluzenter Farbe und TPE (Thermoplastische Elastomere) unterstreichen den hochwertigen und »cleanen« Eindruck der Zahnbürste. Ein markantes und zugleich funktionales Element ist der ovale Durchbruch mit farblicher TPE-Umrandung. Er bildet eine rutschfeste Auflage für den Daumen.

Dieses zentrale Gestaltungsmerkmal findet sich auch in Kopf und Griff wieder. Die farbige Rückseite des Bürstenkopfs hebt die markante Form des Ovals nochmals hervor und setzt einen zusätzlichen Farbakzent.

Die Gestaltung des Bürstengriffes garantiert eine gute Handhabung. So liegt der Griff bei den verschiedensten Positionen des Zähneputzens gleichbleibend angenehm in der Hand.

Die drei Zahnbürsten besitzen unterschiedliche Borstenkonfigurationen, die auf die verschiedenen Anwendungsbereiche ausgerichtet sind und sich in den Farben unterscheiden. Die hellblaue – Sensitive – hat weiche, sanft abgerundete Borsten. Die mintgrüne Bürste – Orthodontic – ist mit einem speziellen Borstenprofil für schwer zugängliche Stellen ausgestattet, während die orange – Interdental – abgerundete Borsten mit einem flachen Profil besitzt, um die Zahnzwischenräume gründlich zu reinigen.

These toothbrushes are designed especially for medical applications. The high standards this involves are reflected in the severe, unassuming design and technoid shape. The toothbrushes have smooth, slim lines and slightly curved forms.

Crystal-clear PET in combination with a sparing use of translucent colour and thermoplastic elastomers (TPE) underscore the toothbrushes' high-quality, clean impression. The oval aperture in the shaft, with its coloured TPE edge, is a striking and at the same time functional element. It also prevents the thumb from slipping.

This central design element is repeated in the head and handle. The coloured rear surface of the brush head once again takes up the distinctive form of the oval, and sets an additional colourful accent.

The design of the brush handle makes for an easy grip. It lies pleasantly in the hand at all times, no matter what teeth-cleaning position has been assumed.

The three toothbrushes have different bristle configurations designed for different areas of application. They also differ in colour. The light-blue, sensitive brush has soft, gently rounded bristles. The mint-green, orthodontic brush is furnished with a special bristle profile for poorly accessible areas, while the orange, interdental brush has rounded bristles with a flat section, to allow the spaces between the teeth to be cleaned thoroughly.

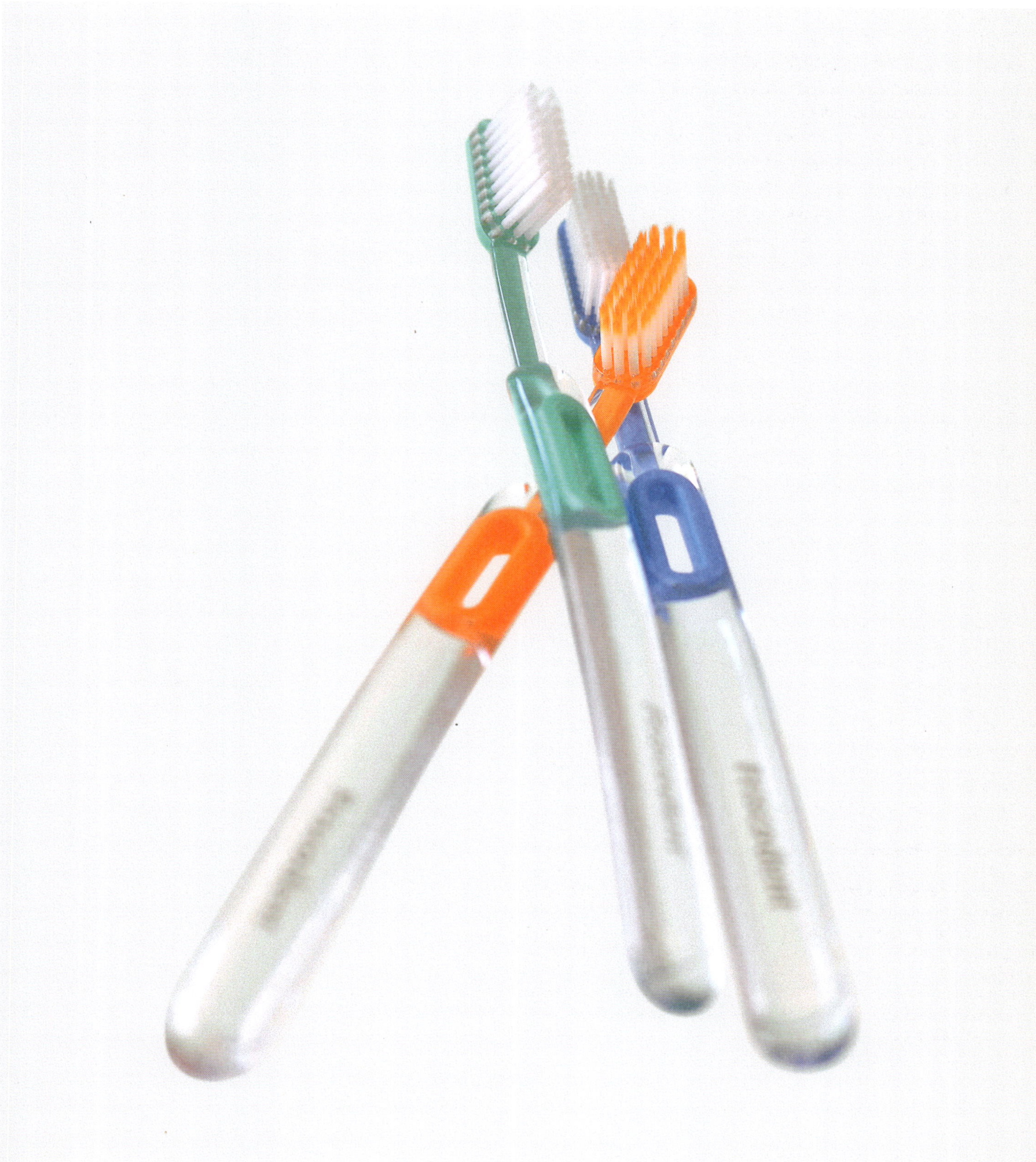

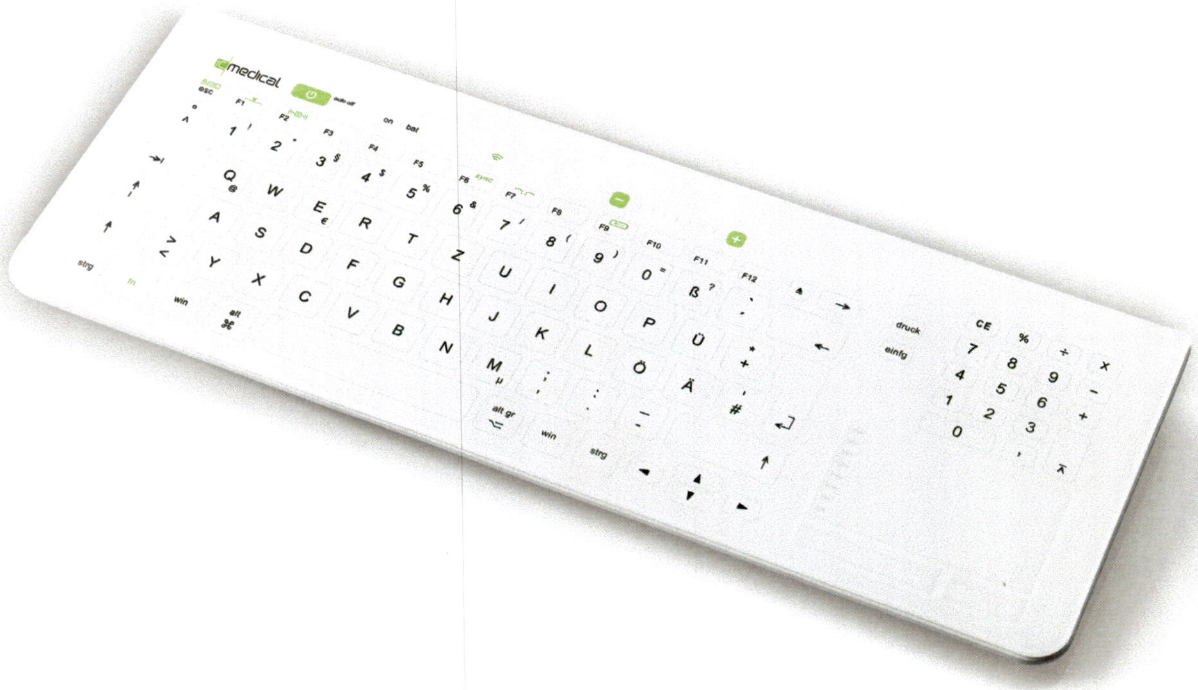
medical
druck
einfg
CE

IC Keys
Glastastatur

Glass keyboard

Medizin,
Rehabilitation, Geriatrie
Silber

Medicine,
rehabilitation, geriatric medicine
Silver

Jury
Judges panel

Eine formal sehr schöne Tastatur, die auch extrem funktional ist. Im Klinikbereich benutzt man normalerweise herkömmliche Tastaturen, die oft nur mit einer Folie überzogen und nicht sehr hygienisch sind. Diese Tastatur ist anders. Sie lässt sich leicht bedienen und ihr Material, Glas, ist extrem gut zu reinigen. Auch die Typografie ist sehr anspruchsvoll, gerade für den medizinischen Bereich. Die Tastatur liegt deutlich über dem Standard.

Formally, a very elegant keyboard, which is also extremely functional. In hospitals, conventional keyboards are still commonly used. Often, they are only covered with a film, which is not very hygienic. This keyboard is different. It is easy to use and its material, glass, is extremely easy to clean. We also think its typography is very upmarket, especially for the medical sphere. The keyboard is clearly a cut about the norm.

Hersteller
Manufacturer
LNT Automation GmbH
D-71397 Leutenbach-Nellmersbach

Design
Designer
Jacob Bachmayer
D-70191 Stuttgart

Vertrieb
Distributor
IC Medical Vertriebsgesellschaft VG
D-73650 Winterbach

IC-Keys entspricht vollkommen der Empfehlung des Robert-Koch-Instituts für Krankenhäuser und Arztpraxen. Dort werden Tastaturen mit »glatten, flüssigkeitsdichten und desinfizierbaren Oberflächen« empfohlen. Die Realität sieht aber noch anders aus. Meist werden Standardtastaturen benutzt, die auf Grund ihres mechanischen Aufbaus nicht zu desinfizieren sind, zwar manchmal mit Folien bedeckt werden, die aber keinen wirklichen Schutz bieten.

Ganz anders dagegen bei IC-Keys. Diese Tastatur mit Touchpad ist wasser- und staubdicht. Dank ihrer geschliffenen Glasfront ist sie schnell und sicher zu desinfizieren. Direkt auf der Tastatur wurde ein großes Trackpad integriert, um eine mögliche Kreuzkontamination mit einer PC-Maus zu vermeiden.

Selbst mit Handschuhen kann auf der Tastatur geschrieben werden. Da die Glasplatte den Fingern beim Berühren kein Feedback liefert, wird dies akustisch gelöst. Jeder Anschlag wird durch einen Knackton bestätigt, der sich in der Lautstärke regeln lässt.

Neben der Funkverbindung kann die Tastatur auch über ein USB-Kabel mit dem PC verbunden werden. Beim Wireless-Betrieb erhält sie ihren Strom über vier AA-Zellen – Batterien oder Akkus, die eine lange Betriebszeit von circa 60 Stunden garantieren.

Die außergewöhnliche Tastatur ist sehr funktional und gleichzeitig formschön. Sie besticht nicht nur durch ihr extrem flaches und ergonomisches Design, sondern auch durch ihre übersichtliche, sehr aufgeräumte und klare Gestaltung der Bedienoberfläche, die für hohe Nutzerfreundlichkeit sorgt.

IC Keys is the perfect response to the recommendations of the Robert Koch Institute for hospitals and surgeries. The institute calls for keyboards with flat, waterproof surfaces that can be disinfected. But things look different in the real world. Usually, standard keyboards are used. Their mechanical construction means they cannot be disinfected, and while they are sometimes covered with film, this cannot offer proper protection.

With IC Keys, it is a different matter. This keyboard with its touchpad is watertight and dustproof. Thanks to its polished glass surface, it can be disinfected quickly and reliably. A large trackpad has also been integrated directly into the keyboard to avoid any potential cross-contamination from a mouse.

The keyboard can still be used while wearing gloves. The problem that the glass surface does not give the fingers any feedback when it is touched has been solved acoustically. Each stroke is confirmed by a clicking sound, the volume of which can be controlled.

The keyboard can be connected to the PC wirelessly or via a USB cable. In wireless operation, it gets its power from 4 AA battery cells or rechargeable batteries, which guarantee a long operating period of about 60 hours.

This unusual keyboard is very functional and at the same time elegant. It is striking not only for its extremely flat and ergonomic design, but also for the clear and very tidy design of the user interface. This makes it very user-friendly.

IC Sense
Wandhygienespender

Wall-mounted hygiene dispenser

Medizin, Rehabilitation, Geriatrie Silber

Medicine, rehabilitation, geriatric medicine Silver

Jury
Judges panel

Sehr schöner, klar und hochwertig gestalteter Wandhygienespender, der sich deutlich von den üblichen Produkten, meist unschönen Kunststoffflaschen, abhebt. Hier dagegen: eine Glasfront in zwei Varianten vor einer Edelstahlbox. Der Spender funktioniert berührungslos und lässt sich durch die Materialwahl gut desinfizieren. Damit erfüllt er hohe Hygienestandards.

Very elegant, clear and outstandingly designed wall-mounted dispenser, which clearly stands out from the usual products, which are generally ugly plastic bottles. What we have here, by contrast, is glass surfaces in two variants, with a stainless steel box underneath. The dispenser works without being touched, and the materials used mean it can be easily disinfected. It therefore fulfils high hygiene standards.

Hersteller
Manufacturer
IC Medical Vertriebsgesellschaft VG
D-73650 Winterbach

Design
Designer
Jacob Bachmayer
D-70191 Stuttgart

Vertrieb
Distributor
IC Medical Vertriebsgesellschaft VG
D-73650 Winterbach

Speziell für kleinere Arzt- und Zahnarztpraxen wurde dieser Wandhygienespender konzipiert. Er unterscheidet sich deutlich von herkömmlichen Modellen. So besteht er nicht nur aus hochwertigen Materialien, sondern arbeitet auch berührungslos. Zudem können Desinfektionsmittelflaschen aller Hersteller eingesetzt werden.

Hygiene ist das oberste Gebot. Deshalb bestehen die Fronten aus gehärtetem Glas – ein Material, das ideal zum Desinfizieren ist. Dabei kann zwischen zwei Varianten gewählt werden: gespiegelt oder weiß hinterdruckt. Das bietet einen zusätzlichen Vorteil, weil in Arztpraxen oft zwei Spender kombiniert werden – einer mit Seife und der andere mit Desinfektionsmittel –, die sich durch die unterschiedlichen Fronten eindeutig zuordnen lassen. Ein Infrarot-Näherungssensor erkennt, wann sich eine Hand unter dem Hahn befindet, und erst dann spendet IC Sense Seife oder Desinfektionsmittel.

Der Korpus und die Mechanik sind aus hochwertigem Edelstahl gefertigt, das wie Glas den Ansprüchen absoluter Sauberkeit entspricht. Die Materialkombination ermöglicht eine sehr kompakte und dezente Bauform. Damit ist der Spender um ein Drittel kleiner als alle Konkurrenzprodukte und fügt sich so harmonisch auch in kleine Behandlungsräume ein.

Die Stromversorgung erfolgt über vier AA-Zellen – Batterien oder Akkus. Zwei LEDs zeigen den Energiestatus an.

IC Sense hebt sich sowohl durch hohe Funktionalität und hochwertigen Materialeinsatz als auch durch seine elegante, reduzierte und sachliche Formgebung von handelsüblichen Wandhygienespendern ab.

This wall-mounted hygiene dispenser has been designed especially for doctors' and dentists' surgeries. It stands clearly apart from conventional models. Not only is it made of high-quality materials, it also works without being touched. Disinfectant bottles of all makes can be inserted into it.

The number-one priority is hygiene. This is why its surfaces are made of hardened glass – a material that is ideal for disinfecting. There is a choice of two variants: mirror glass or glass on a white backing foil. The additional benefit of this is that surgeries often have two kinds of dispenser – one for soap and one for disinfectant. Each can then be allocated to one of these variants. Only when an infrared proximity sensor detects a hand under the tap does IC Sense dispense soap or disinfectant.

The body and mechanism are made of high-quality stainless steel. Like glass, this satisfies the demands of absolute cleanliness. The combination of materials also allows the dispenser to be very compact and unobtrusive. As a result, the dispenser is roughly a third smaller than all competing products, and blends in harmoniously even in small treatment rooms.

It gets its power from four AA battery cells or rechargeable batteries. Two LEDS show the level of charge.

A high level of functionality, high-quality materials, and an elegant and sober design make IC Sense stand out from the commercially available wall-mounted dispensers.

medical

E-scope®
Ophthalmoskop / Otoskop

Ophthalmoscope / otoscope

Medizin,
Rehabilitation, Geriatrie
Silber

Medicine,
rehabilitation, geriatric medicine
Silver

Jury
Judges panel

Dieses Instrument bietet eine revolutionäre Innovation: Es ist mit normalen AA-Batterien verwendbar und arbeitet trotzdem mit LED-Technik. Diese ungewöhnliche Kombination ermöglicht sehr helles Licht. Es ist einhändig zu bedienen und unterscheidet sich sowohl formal als auch durch seine Materialität deutlich von den Geräten, die es momentan auf dem Markt gibt. Es wirkt freundlicher und angenehmer, was vielleicht auch dem Patienten ein bisschen die Furcht nimmt.

This instrument presents a revolutionary innovation: it can be used with normal AA batteries, but is nonetheless equipped with LED technology. As a result of this unusual combination, it is very bright. It can be operated single-handedly. Formally and in terms of the materials used, it differs significantly from the devices currently in the market. It is friendlier and more pleasant in appearance, which may also help to allay patients' fears a little.

Hersteller
Manufacturer
Rudolf Riester GmbH
D-72417 Jungingen

Design
Designer
RED GmbH
D-70182 Stuttgart

Vertrieb
Distributor
Rudolf Riester GmbH
D-72417 Jungingen

E-Scope® verkörpert eine Weltneuheit. War die Verbindung von LED-Beleuchtung und handelsüblichen Alkaline-Batterien bisher nahezu unmöglich, so ist es mit diesem Tascheninstrument gelungen: die ideale Kombination aus hochmoderner LED-Technologie, effizienter Diagnose und umweltschonender Energiequelle.

Möglich ist dies dank dem neuentwickelten IPC (Integrated Power Converter). Dieser wandelt die Spannung einfacher Alkaline Batterien in die für den LED-Betrieb stets optimale Spannung um und ermöglicht Batterielaufzeiten von rund 50 Betriebsstunden.

Die Vorteile liegen klar auf der Hand: geringe Unterhaltskosten durch lange LED- und Batterielebensdauer. Beträgt die Lebensdauer einer bisher üblichen Xenon-Lichtquelle nur circa 20 Stunden, so leuchtet die LED mindestens 20.000 Stunden, also über zwei Jahre am Stück.

Das reinweiße Licht sorgt für eine farbneutrale Ausleuchtung sowie eine bessere Farbdifferenzierung. Dadurch unterstützt es eine effizientere Diagnose im Vergleich zu Halogen- oder Xenonlampen.

Das macht das E-Scope® zum wegweisenden Tascheninstrument für die flexible Diagnose. Die LED-Technik wurde hier integriert in ein Ophthalmoskop – ein Instrument zur Augenspiegelung – und in ein Otoskop, mit dem der Arzt oder der Hörgeräteakustiker den äußeren Gehörgang einschließlich des Trommelfells untersucht. Beide Instrumente, die es in Schwarz und Weiß gibt, sind auch mit Vakuum- oder Halogen-Beleuchtung erhältlich.

E-Scope® is a world first. While combining LED illumination and commercially available alkaline batteries has so far been practically impossible, this pocket instrument has succeeded in doing so. It is an ideal combination of state-of-the-art LED technology, efficient diagnosis and eco-friendly power source.

This is possible thanks to the newly developed integrated power converter, or IPC. It converts the voltage of simple alkaline batteries into the optimum constant voltage for LED operation, and allows batteries to last for roughly 50 hours of operation.

The benefits are obvious: low maintenance costs as a result of long LED and battery life. While the service life of the xenon light sources used up to now is only about 20 hours, the LED will work for at least 20,000 hours, or for two years without interruption.

Its pure white light ensures colour-neutral illumination and better colour differentiation. In this way, it helps make diagnosis more efficient than with halogen or xenon light sources.

All this makes the E-Scope® a groundbreaking pocket instrument for flexible diagnosis. In this instrument, LED technology has been integrated into an ophthalmoscope for examining the eyes and an otoscope for doctors or hearing-aid audiologists examining the external auditory canal, including the ear drum. Both instruments, which are available in black and white, are also available with vacuum or halogen illumination.

e-scope®
Riester
e-scope®

Scala
Laboreinrichtung

Laboratory workplace

Jury
Judges panel

Medizin, Rehabilitation, Geriatrie
Silber

Medicine, rehabilitation, geriatric medicine
Silver

Endlich mal ein modernes modulares System, das extrem funktional ist, ausgezeichnete Produktmerkmale bietet und jeden Menschen, der im Labor arbeitet, begeistern wird. Laborarbeit ist hochpräzise – je weniger Ablenkung da ist, desto besser kann man sich auf seine Arbeit konzentrieren.

At last, a modern, modular system that is extremely functional, offers excellent product characteristics, and will excite anyone who works in a laboratory. Laboratory work has to be extremely precise. The less distraction people have, the better they will be able to concentrate on their work.

Hersteller
Manufacturer
Waldner Laboreinrichtungen
GmbH & Co. KG
D-88239 Wangen

Design
Designer
Idea Design Team
D-88239 Wangen

Vertrieb
Distributor
Waldner Laboreinrichtungen
GmbH & Co. KG
D-88239 Wangen

Das neue Laboreinrichtungsprogramm Scala besticht nicht nur durch sein ästhetisches, klares und zeitloses Design – es bringt vielmehr innovative, hochfunktionale Lösungen für die unterschiedlichsten Arbeiten im Labor.

Zentrales Element ist die Medienversorgung durch den Kanal in der Medienzeile, der die für Laborarbeiten typischen Medien bereitstellt, wie 220/380 Volt, EDV-Daten, verschiedene Wasserqualitäten und diverse Gase.

Die markante Neigung des Medienkanals bringt Armaturen, Installationen und Ablageflächen näher zum Nutzer und gleichzeitig bietet sie dadurch eine größere nutzbare Tiefe der Arbeitsfläche.

Für eine optimale Orientierung im Labor sorgt die farbliche Gestaltung, die die unterschiedlichen Arbeitsbereiche und Ebenen klar definiert. Dabei unterstützt das helle Grau die zurückhaltende kontrastarme Arbeitsumgebung im Labor, sodass Bedienungselemente wie Armaturen und Schalter durch ihre jeweilige Farbe gut zu erkennen und sicher zu bedienen sind.

Im Labor muss oft an der gleichen Stelle stehend und sitzend gearbeitet werden. Dafür wurden eigens Schiebeelemente entwickelt – eine Innovation in der Laboreinrichtung, wie beispielsweise der Splitter- und Spritzschutz »Protector«. Oder der »Sekretär«, an dem man stehend schreiben oder in den Laptop eingeben kann, während der »Assistent« eine abklappbare Tischfläche für sitzende Tätigkeiten bietet. Diese verschiebbaren Module erhöhen die Funktionalität, Ergonomie und Sicherheit am Laborarbeitsplatz deutlich.

This new laboratory furnishing system stands out not only for its tasteful, clear and timeless design, but also because it offers innovative and highly functional solutions for a wide variety of laboratory tasks.

Its central element is a media supply via a channel in the media row. It supplies the typical media needed for laboratory work, such as a 220/380 volt power supply, IT data, various water qualities, and diverse gases.

The strikingly angled media channel brings fittings, equipment and shelves closer to the user, and at the same time increases the workspace's usable surface.

Optimum orientation in the laboratory is provided by a colour scheme that clearly defines different work areas and levels. The light grey colour supports the unassuming, low-contrast working environment in the laboratory, making the different colours of the operating elements such as fittings and switches easy to recognize and safe to operate.

People working in a laboratory often have to switch from sitting to standing at the same workplace. Moveable elements have been designed especially for this. One example of this innovation in laboratory furniture is the »Protector« splinter and spray guard. Or there is the »Sekretär«, at which people can write or use a laptop while standing, and the »Assistent«, which provides a fold-out desk top for sedentary activities. These moveable elements significantly improve functionality, ergonomics and safety at the laboratory workplace.

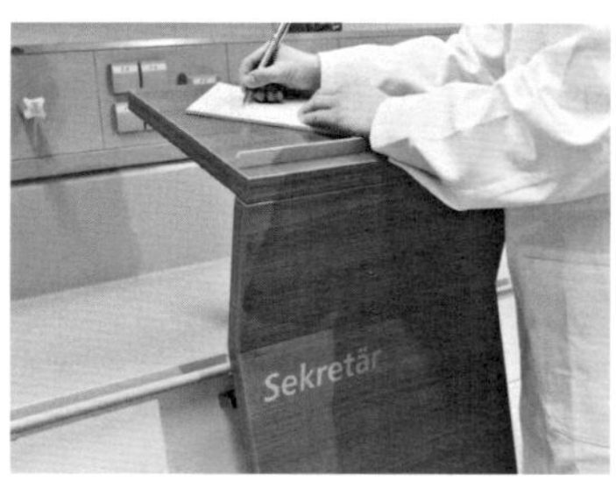

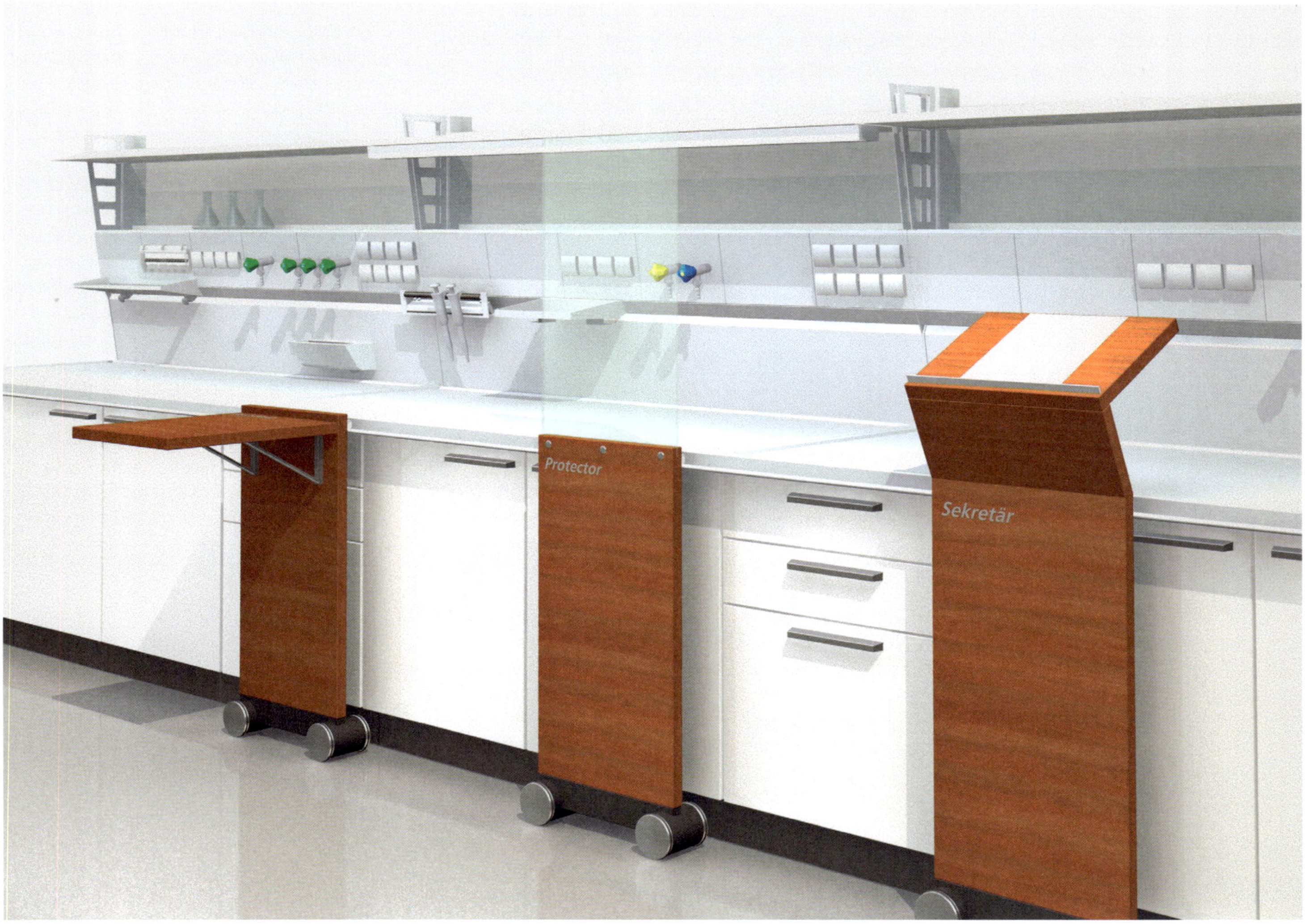
Protector
Sekretär

Jury
Judges panel

Alegro
Monitorträgersystem

Medizin,
Rehabilitation, Geriatrie
Silber

Monitor carrier system

Medicine,
rehabilitation, geriatric medicine
Silver

Eine gestalterisch wie funktional hervorragend gelöste Monitoranbindung mit ultraflexiblen Einsatzmöglichkeiten. Der Hersteller hat in jedem Detail über die spätere Anwendung des Endbenutzers nachgedacht und sich Produktionsschwierigkeiten auferlegt zugunsten einer sehr komfortablen Bedienung. Das Trägersystem lässt sich mit einem Handgriff bedienen.

Design-wise and functionally, an excellent solution for attaching a monitor, with extremely flexible possibilities for use. In every detail, the manufacturer has thought about later use by the final customer, and has taken on challenges in production in order to make it easy to operate. The carrier system can be used blindfolded with just one hand movement.

Hersteller
Manufacturer
Haeberle GmbH & Co. KG
D-70565 Stuttgart

Design
Designer
Fuchs Design
D-70599 Stuttgart

Vertrieb
Distributor
Haeberle GmbH & Co. KG
D-70565 Stuttgart

Das Monitorträgersystem Alegro beeindruckt durch sein klares Design und seine hohe Funktionalität sowie durch seine Vielseitigkeit. Es ermöglicht, Monitore oder Panel-PCs sowohl einzeln als auch mehrfach anzuordnen – neben- oder übereinander, an der Wand, auf dem Schwenk- oder Gelenkarm oder an speziellen Profilen. Über zwei Einstellhebel lässt sich Alegro einfach horizontal und vertikal drehen, neigen und schwenken.

Durch seine unkomplizierte und universelle Positionierungsvielfalt kann er für ganz verschiedene Anwendungen nicht nur in der Medizin, sondern auch in der Industrie und in sämtlichen technischen Bereichen eingesetzt werden.

Und er lässt sich einfach montieren. Die Kabel werden optimal innerhalb des Trägerprofils, in den Gelenkarmen oder mit Kabelclips entlang der Schwenkarme geführt.

Die flexible Regulierung des Monitorträgersystems ermöglicht eine perfekte Anpassung an die persönlichen Arbeitsanforderungen und erfüllt dabei hohe Ansprüche in puncto Funktion, Handhabung und Ergonomie.

Dabei zeichnet sich Alegro durch eine ausgeprägte Stabilität und hohen Bedienkomfort aus.

The Alegro monitor carrier system is impressive for its clear design and good functionality, as well as for its versatility. It allows monitors or panel PCs to be arranged singly or in multiples, next to each other or on top of each other, on the wall, on a swivel arm or articulated arm, or on special sections. Two levers allow Alegro to be turned horizontally and vertically, as well as to be tilted and swivelled.

This uncomplicated variety of positions means it can be used for many different applications: not only in medicine, but also in industry and technical areas of all kinds.

And it is easy to mount. Cables run inside the load-bearing section or articulated arms, or are guided along the swivel arms by means of cable clips.

The flexible configurations of the monitor carrier system allow it to be matched perfectly to personal working requirements. It meets high standards of function, handling and ergonomics.

Alegro is also exceptionally robust and easy to operate.

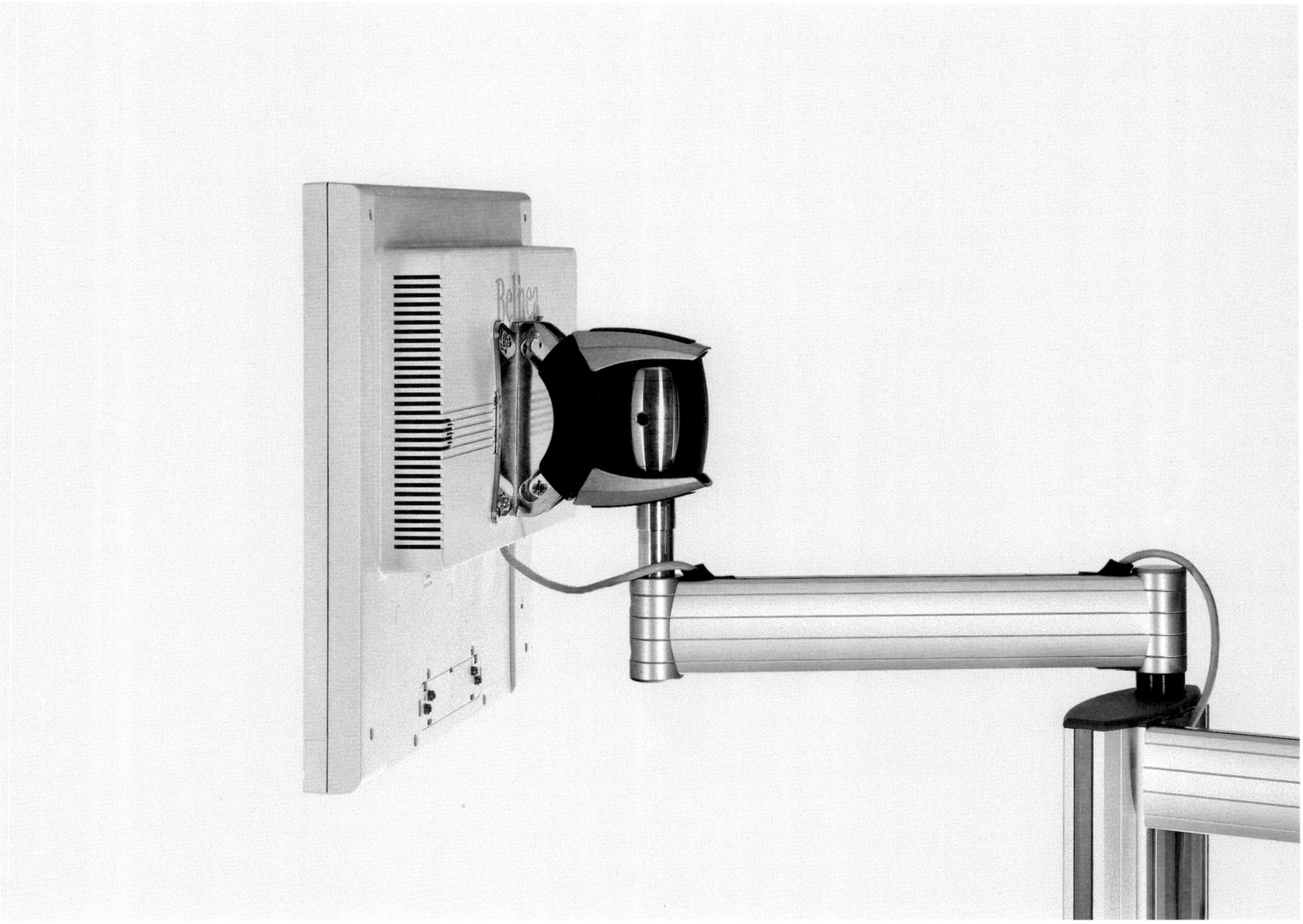

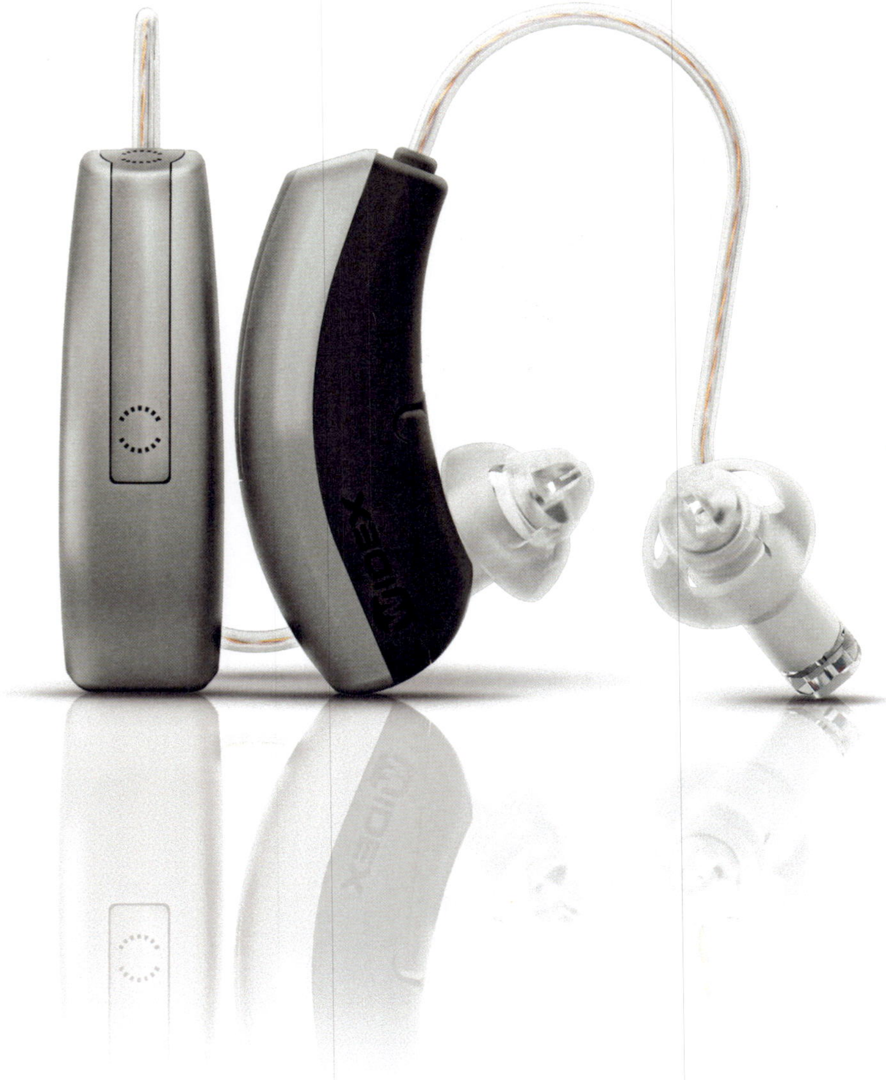
widex

Widex clear440 mit Widex Dex
Hörgerät und Zubehör

Hearing aid and accessories

Medizin,
Rehabilitation, Geriatrie
Silber

Medicine,
rehabilitation, geriatric medicine
Silver

Jury
Judges panel

Das Hörgerät ist klein, handlich, unauffällig, nicht diskriminierend, modern, funktional und schön gestaltet. Man kann verschiedene Modi ganz einfach einstellen, zum Telefonieren, Fernsehen oder Musikhören. Der ausgesprochen einfache Batteriewechsel überzeugt, weil gerade dies für ältere Menschen oft sehr schwierig ist. Es ist ein herausragendes Beispiel dafür, dass diese Art Produkte im Design mehr Beachtung finden. Dieses Produkt setzt in der Branche neue Impulse.

The hearing aid is small, handy, inconspicuous, non-discriminating, modern, functional, and beautifully designed. Very simply, various modes can be set, such as telephoning, watching television, or listening to music. It is good to see that it is so easy to change the batteries, since this is often very difficult for the elderly. It is an excellent example of how products such as these are being given more attention by designers. It sets a new trend in the industry.

Hersteller
Manufacturer
Widex Hörgeräte GmbH
D-70567 Stuttgart

Design
Designer
Widex A/S
DK-3540 Lynge

Vertrieb
Distributor
Widex Hörgeräte GmbH
D-70567 Stuttgart

Eine Weltneuheit ist das Hörsystem Widex clear440. Mit Hilfe der revolutionären C-ISP Kommunikationsplattform ist es gelungen, die wichtigsten Eigenschaften des natürlichen Hörens nachzubilden. Das ermöglicht erstmals räumliches Hören mit dreidimensionaler Orientierungsfähigkeit.

Mit der innovativen Zubehörlinie Widex Dex bietet das Hörsystem außerdem eine einzigartige Lösung für drahtloses Fernsehen, Musikhören und mobiles Telefonieren: erstmals in Echtzeit, in Stereo-Hi-Fi-Qualität und beispiellos einfach zu bedienen. Dafür wurde eine eigene Funktechnologie entwickelt, die den bisherigen Bluetooth-Standards weit überlegen ist.

Widex clear440 besticht durch herausragende Gestaltung und hohe Klangbrillanz, optimales Sprachverstehen und sogar durch Vorteile gegenüber Normalhörenden.

Mit der C-ISP Kommunikationsplattform wurde eine völlig neue Chiparchitektur entwickelt. Sie lässt die zwei Hörgeräte drahtlos miteinander kommunizieren und ermöglicht einen exakten »Fingerabdruck« des Ohrs für maßgeschneiderte Hörlösungen.

Widex clear440 setzt nicht nur durch seine einzigartige Technologie einen Meilenstein, sondern auch als weltweit kleinstes Hörsystem; denn es ist gelungen, die Technologie in das nahezu unsichtbare Designgehäuse zu integrieren.

Mit dem neuen Hörsystem wurden Kommunikationslösungen geschaffen, die Menschen mit Hörminderung den Zugang zur modernen Kommunikationswelt in einer bahnbrechenden Qualität ermöglichen.

The Widex clear 440 hearing-aid system is a world first. With the help of the revolutionary C-ISP communication platform, Widex has succeeded in reproducing the qualities of natural hearing. For the first time, spatial hearing is possible, with three-dimensional orientation.

Together with its innovative Widex Dex range of accessories, this system also offers a unique solution for wireless television, listening to music and mobile telephony: for the first time ever in real time and stereo-hifi quality, and easy to operate at the same time. To achieve this, the company developed its own radio technology, which is far superior to the bluetooth standards used up to now.

Widex clear440 stands out by virtue of its excellent design and acoustic brilliance. Speech can be understood perfectly, and understanding is even better than for people with normal hearing.

The C-ISP communication platform is based on a completely new chip architecture. This allows two hearing aids to communicate with each other wirelessly, and allows a precise »fingerprint« of the ear to be taken, for customized hearing solutions.

But it is not only this exceptional technology that sets a milestone. Widex clear440 is also the world's smallest hearing aid, with all this technology integrated into the almost invisible designer housing.

This new hearing aid system creates communication solutions that allow people with impaired hearing to access the modern world of communications, and this in groundbreaking quality.

HB 1994C
Gipsschere

Plaster cast cutter

Medizin,
Rehabilitation, Geriatrie
Silber

Medicine,
rehabilitation, geriatric medicine
Silver

Jury
Judges panel

Ein puristisches Produkt, die Schere in einer wunderbaren Materialhaptik, ausgesprochen sauber und hochwertig gestaltet, solide und ergonomisch. Von der Wertanmutung sehr schön.

A puristic product, wonderful haptic qualities of the cutter's materials, exceptionally neat and high-quality design, robust and ergonomic. Very appealing, high-quality air.

Hersteller
Manufacturer
Hebumedical GmbH
D-78532 Tuttlingen

Design
Designer
Martin Storz Design
D-78532 Tuttlingen

Vertrieb
Distributor
Hebumedical GmbH
D-78532 Tuttlingen

Im Gegensatz zu herkömmlichen Modellen berücksichtigt diese Gipsschere bei der Gestaltung die Anatomie der Hand. Daraus ergibt sich eine asymmetrische Formgebung, die eine stabilere Verbindung zwischen Hand und Instrument ermöglicht. Die Hand hält die Schere sicher und gleitet nicht mehr ab wie bei den bisherigen glatten Griffen.

Auf der einen Griffseite sind vier Fingerkuhlen integriert, deren Anordnung die unterschiedliche Länge der Finger berücksichtigt. Auf der anderen Seite dient eine konkave Mulde als Bett für den Daumenballen. Zusätzlich wurde die Griffweite zwischen Daumen und Zeigefinger so vergrößert, dass sie eine optimale Kraftübertragung garantiert.

Das neue ergonomische Griffdesign garantiert einen sicheren Halt, einen verstärkten Kontakt zwischen Hand und Instrument und dadurch eine verbesserte Führungsstabilität. Gleichzeitig ist der Druck auf die spannungsfrei eingesetzten Finger erheblich reduziert.

Die ballig geformte Scherenspitze ist innenseitig kantenfrei gestaltet, um Verletzungen am Patienten zu verhindern.

Unlike conventional models, the design of this plaster cast cutter takes the anatomy of the hand into account. The result is an asymmetrical shape that allows a more stable connection between the hand and the instrument. The cutter is held firmly in the hand and no longer slides off, as is the case with the smooth grips available up to now.

On the one side, there are four depressions for the fingers, arranged to allow for the fingers' different lengths. On the other side, a concave hollow serves as a bed for the ball of the thumb. In addition, the width of the grip between the thumb and forefinger has been increased to ensure an optimum transfer of force.

The new ergonomic design of the grip guarantees that it is grasped firmly. Contact between the hand and the instrument is improved, and with it the stability of the cut. At the same time, there is considerably less pressure on the other fingers.

The spherically formed tip of the cutter has no sharp edges on the inside, to keep patients from being injured.

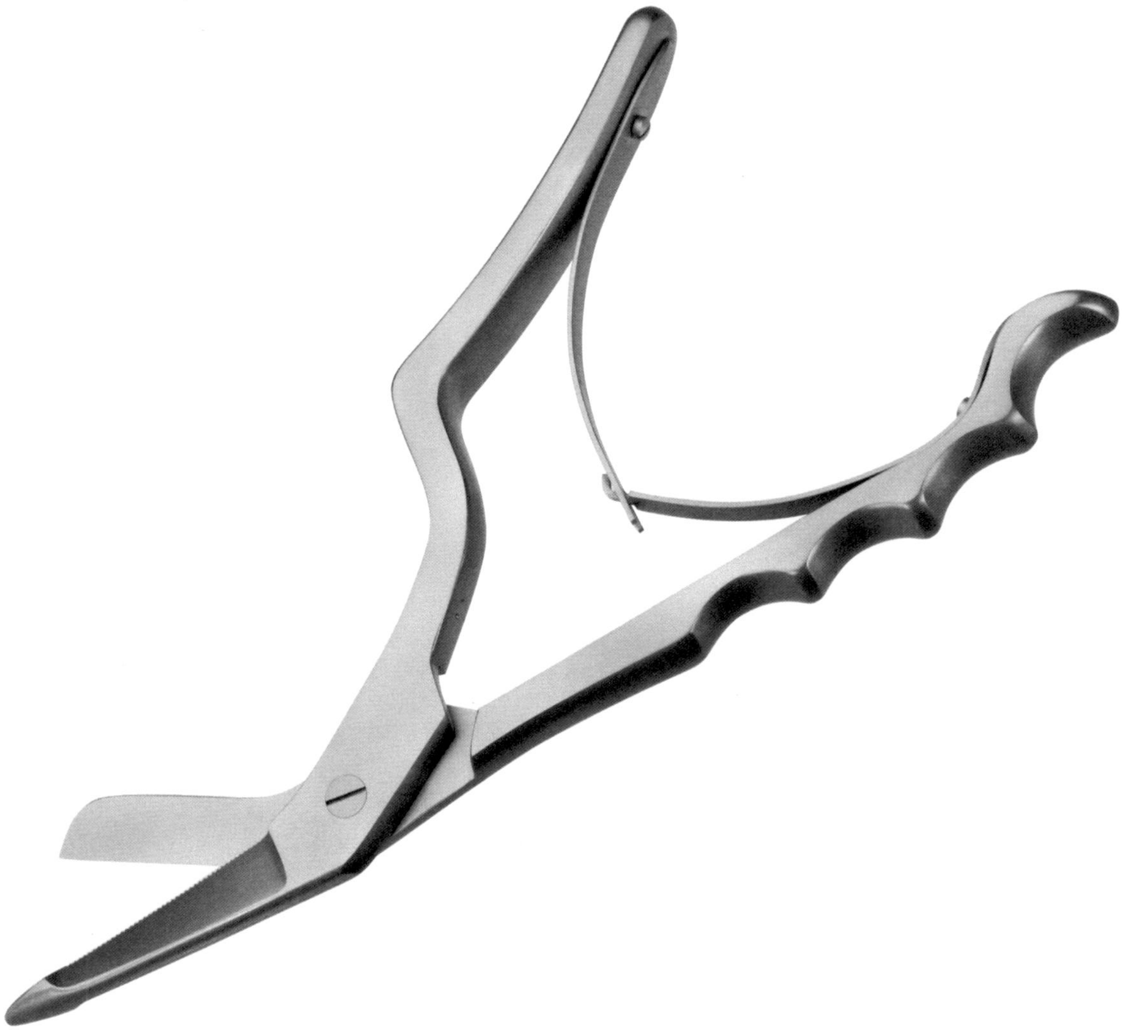

P2
Inkubatorplattform

Incubator platform

Medizin, Rehabilitation, Geriatrie Silber

Medicine, rehabilitation, geriatric medicine Silver

Jury
Judges panel

Schöne, wertanmutige Gestaltung, die weit über das in der Medizintechnik Übliche hinausgeht. Hohe Funktionalität und Stabilität bei minimalem Materialeinsatz, weil es aus dem Vollen gefräst wurde, wie im Flugzeugbau üblich. Sie ist modular aufgebaut und leicht, man kann die Teile jederzeit austauschen. Durch die niedrige Bauhöhe passt sie auch in neue Hubschrauber. Gute Montage- und Verkabelungsmöglichkeiten. Ein Produkt, das eine absolute Zuverlässigkeit ausstrahlt.

Elegant, high-quality design that goes far beyond what is usual in medical technology. High level of functionality and stability using a minimum of materials. It has been milled from a solid block of aluminium – a common method in aircraft building. It is modular and light, the parts can be exchanged at any time. Its low height means it fits into modern helicopters as well. Attaching appliances and cables is easy. A product that emanates absolute reliability.

Hersteller
Manufacturer
Starmed GmbH
D-89081 Ulm

Design
Designer
Werksdesign
In-house design
René Stern

Vertrieb
Distributor
Hissin Medizintechnik GmbH
D-61440 Oberursel / TS

Für den Transport von Früh- oder Neugeborenen von Klinik zu Klinik benötigt man eine Plattform, die mit einem Baby-Transportinkubator und Medizingeräten bestückt wird. So kann das Baby beatmet und umfangreich überwacht werden.

Herkömmliche Plattformen sind meist optisch wenig ansprechend und wenig innovativ, unflexibel und sehr schwer. Kabel und Schläuche der medizinischen Geräte lassen sich oft nur schlecht unterbringen, und durch einen zu hohen Aufbau können sie in neuen Hubschraubern nicht eingesetzt werden.

Herausgekommen ist eine leichte, schlichte und formschöne Inkubatorplattform, die sich deutlich von den herkömmlichen Modellen unterscheidet. Sie zeichnet sich durch ein flexibles Baukastensystem und ein neuartiges Konstruktionsprinzip aus. Die Räder wurden an die Plattformenden verlagert, wodurch die Bauweise nun niedriger ist. Durch die taillierte Bauform erreichte man eine geringere Breite. Die Anzahl der Bauteile und das Gewicht konnten reduziert werden.

Die Plattform besitzt nun einen integrierten Kabelkanal und universell nutzbare Befestigungsmöglichkeiten für unterschiedlichste Adapter. Dadurch ist eine schnelle Montage und Demontage von Transportinkubator und Medizingeräten möglich. Alle Teile der Plattform sind aus Aluminium gefräst und hochwertig verarbeitet.

When transporting premature or new-born babies from one hospital to another, a platform is needed that can be equipped with a travel incubator and medical apparatus. This allows the baby to be supplied with oxygen and monitored closely.

Conventional platforms are usually rather unattractive visually. They are generally not very innovative, inflexible, and very heavy. There is often very little room for the cables and hoses of the medical appliances, and their excess height means they cannot be used in modern helicopters.

These considerations have given rise to a light, unassuming and elegant incubator platform that stands out significantly from conventional models. It is based on a flexible modular system and an innovative design principle. Its wheels have been moved to the platform ends, making the design lower. Tapering the platform has also made it slimmer. It was also possible to reduce the number of parts, and thus the weight.

The platform now has an integrated cable conduit and universal attachments for a broad range of adapters. This allows the transport incubator and the medical appliances to be attached and removed quickly. All parts of the platform have been milled from aluminium and finished to a high standard.

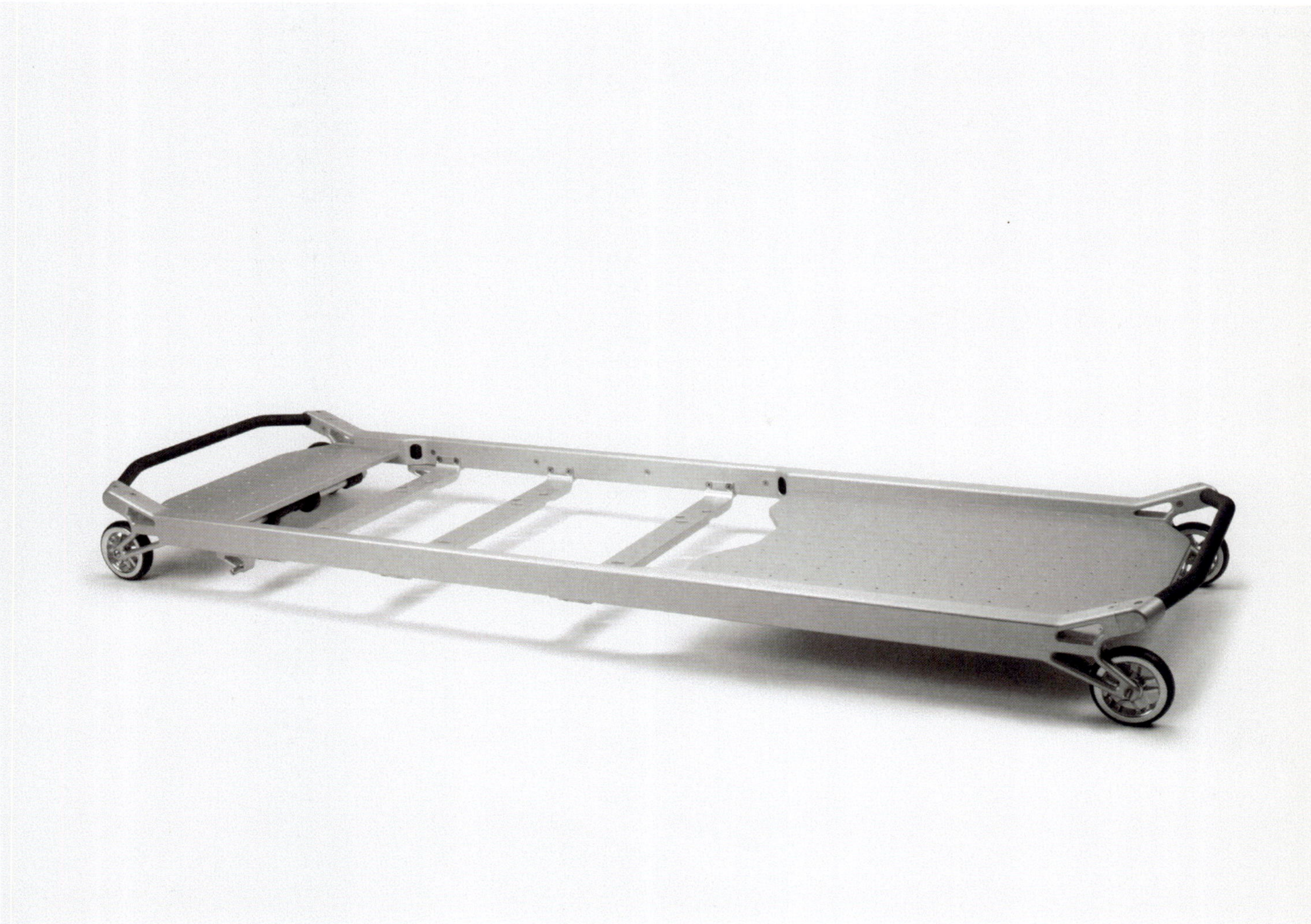

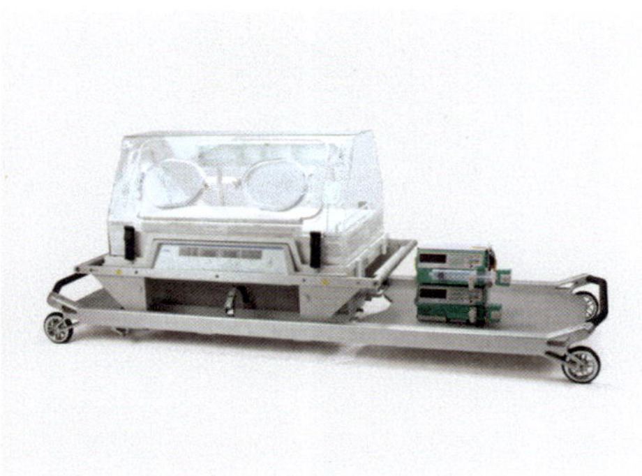

Hersteller
Manufacturer
Duravit AG
D-78132 Hornberg

Design
Designer
EOOS Design GmbH
A-1010 Wien

Vertrieb
Distributor
Duravit AG
D-78132 Hornberg

OPEN SPACE DURAVIT

OpenSpace Duschabtrennung

Bad, Sanitär, Wellness Gold

Shower enclosure

Bathrooms, sanitary installations, wellness Gold

066 067

OpenSpace Duschabtrennung

Shower enclosure

Bad, Sanitär, Wellness Gold

Bathrooms, sanitary installations, wellness Gold

Jury
Judges panel

Ein Produkt, das ein ganz großes Problem im Sanitärbereich hervorragend löst: Die meisten Bäder sind eher klein. Dies ist eine großzügige, voll funktionsfähige Dusche, die man einfach aufräumen und damit Platz gewinnen kann. Genial. Man kann den Raum mit zwei unterschiedlichen Möglichkeiten gestalten. Die verspiegelte Variante bietet einen Zusatznutzen: zum Beispiel kann man sich davor frisieren oder rasieren.

A product that excellently solves a truly big problem in the bathroom area: most people don't have a lot of room in their bathrooms. This is a generous, fully functional shower that I can simply fold away, opening up new space. This is a brilliant idea. The available space can be structured in two different ways. With the mirror-glass variant, you can even do your hair or shave in front of it.

Hersteller
Manufacturer
Duravit AG
D-78132 Hornberg

Design
Designer
EOOS Design GmbH
A-1010 Wien

Vertrieb
Distributor
Duravit AG
D-78132 Hornberg

OpenSpace – die Duschabtrennung zum Wegklappen – vergrößert das Bad und schafft Offenheit im Raum.

Die Kontur an der Wand bildet ein umlaufender Rahmen in hochglänzender Chromoptik. Darin sind zwei großflächige, selbstarretierende Türen aus Glas eingelassen, die entweder transluzent oder einseitig verspiegelt sind. Nach dem Duschen klappt man sie einfach an die Wand.

Keine störende Schiene ragt mehr in den Raum. Armaturen und Brauseschlauch, Duschgels und Shampoos sind versteckt. Diese neue Form der Duschabtrennung öffnet den Raum, vor allem in der Variante mit dem verspiegelten Glas – nicht nur optisch, sondern erweitert den Bewegungsradius im Bad – ein Vorteil besonders auch für kleine Bäder.

Möglich ist dies durch eine technische Besonderheit in den Profilen. Das Türprofil verfügt über einen Hebe- und Senkmechanismus. Dadurch halten die Türen, wenn sie bewegt werden, automatisch bei der Duschposition von 90 Grad an und senken sich. Optimaler Spritzschutz ist so garantiert. Um sie einfacher zu reinigen, können sie noch weiter geöffnet werden und senken sich dann automatisch wieder bei 180 Grad.

Für zusätzliche Stabilität sorgt das integrierte U-Profil. Gleichzeitig ist damit ein Wandausgleich von 18 Millimetern berücksichtigt – ein Vorteil bei nicht ganz geraden Wänden – wie das häufiger im Altbau der Fall ist.

OpenSpace – the shower enclosure that can be folded away – makes the bathroom larger and creates open space.

A frame in brilliant chrome provides the contours on the wall. Two large self-locking glass doors are fitted in the frame, in a translucent design or with a mirror on one side. After showering, these doors can simply be folded against the wall.

There are no more irksome rails protruding into the room. Tap fittings, shower hose, shower gel and shampoo are hidden from view. This new type of shower enclosure opens up the room, especially in its mirror-glass variant. It does so not only optically, but also frees up room for manoeuvre in the bathroom, which is especially advantageous for small bathrooms.

This is made possible by a special technical feature in the door frames. The door frame is fitted with a lifting and lowering mechanism. When the doors are moved, they automatically stop and lower into place at the 90-degree shower position. This ensures the best possible protection against spray. To allow them to be cleaned more easily, they can be opened still further, lowering again automatically when they reach a 180-degree position.

The integrated U section provides for extra stability. At the same time, it means that up to 18 millimetres allowance can be made for the uneven walls that are frequently encountered in older houses.

Axor Bouroullec Kollektion

Collection

Bad, Sanitär, Wellness Silber

Bathrooms, sanitary installations, wellness Silver

Jury
Judges panel

Interessant finden wir die unkonventionelle Art des Spielens mit verschiedenen Ebenen und den Armaturen, die man je nach Wunsch platzieren kann. Sehr überzeugend sind auch die Ablagen, die sonst in der Form nicht vorhanden sind und gut funktionieren, weil sie leicht nach oben gewölbt sind. Die abgelegten Utensilien können so nicht ins Waschbecken abrutschen. Die Form gefällt uns ausgesprochen gut. Das gesamte Konzept ist in seiner Anmutung eine neue Lösung und wirkt durch die Verschachtelung architektonisch. Sehr elegant.

What is interesting for us is the unconventional way of playing with different levels and fittings, which can be located as desired. We are also very taken with the shelves, which are nowhere else to be found in a form like this. They work well, because they are slightly curved upwards at the edges. Nothing that is placed on them can slide into the wash basin. We especially like the form: the entire concept is a new solution, also in its gracefulness. Its blend of different levels gives it an architectural appearance. Very elegant.

Hersteller
Manufacturer
Hansgrohe AG
D-77761 Schiltach

Design
Designer
Ronan und Erwan Bouroullec
F-75010 Paris

Vertrieb
Distributor
Hansgrohe AG
D-77761 Schiltach

Durch ihren modularen Charakter lassen sich mit der Kollektion Axor Bouroullec fast 80 Elemente frei kombinieren. Damit bietet sie bislang einzigartige Möglichkeiten zur individuellen Badgestaltung und eröffnet die Chance herauszufinden, wie man selbst im Bad leben will, welche die jeweils perfekte persönliche Lösung ist.

Kombiniert werden Waschtische, Wanne und Armaturenelemente der Kollektion. Dabei kann die Armatur individuell positioniert werden, denn die in die Waschtische und Wanne integrierten Ablagen können nach Wunsch gebohrt werden. Das ist ohne aufwändige Sonderanfertigung möglich, weil die Waschtische aus Mineralguss gefertigt sind – einem Material, das leicht und problemlos zu bohren ist.

Gestalterisches und funktionales Schlüsselelement sind die Ablagen. Der Nutzer allein entscheidet, wie er sie einsetzen möchte, zum Beispiel als Abstellfläche für Shampoos und Seifen, für Accessoires oder als Platz für den Wasserzulauf und Bedienelemente.

Die Waschtische formen mit ihren Ablagen, die sich auf mehreren Ebenen aus dem Becken erheben, Landschaften im Bad. Ausläufe in verschiedenen Höhen setzen darin markante Akzente. Vom Waschplatz über Dusche und Wanne hinweg entsteht so eine ganz eigene, reizvolle Topografie.

Die durchgängige, sanfte Formensprache, die zurückhaltend, minimalistisch und organisch ist, unterstützt die freie Kombinierbarkeit – und fügt sich harmonisch selbst in bestehende Bad- und Stilwelten ein.

The Axor Bouroullec collection's modular character means that nearly 80 elements can be freely combined. It thus offers previously unheard-of possibilities for individual bathroom design, and gives an opportunity to find out what kind of bathroom is most suitable, and which individual solution is the right one.

The collection's wash stands, bath tub and fitting elements can be combined. The fittings can be positioned individually, since the shelves integrated in the wash stands and tub can be drilled as desired. This is possible without any special customization, because the wash stands are made of mineral cast – a material that can easily be drilled without any problem.

The shelves are the key design and functional element. The user alone decides how he wants to use them – as a space for shampoo and soap, for accessories, or for the water inlet and operating elements.

The wash stand's shelves, which stand out from the bowl on several levels, create bathroom landscapes. Outlets of varying heights allow striking accents to be set. From wash basin to shower and tub, a distinctive and attractive topography is created throughout the entire bathroom.

The flowing, soft formal expression, which is unassuming, minimalistic and organic, supports the freedom to compose at will, and blends harmoniously even into existing bathroom arrangements and styles.

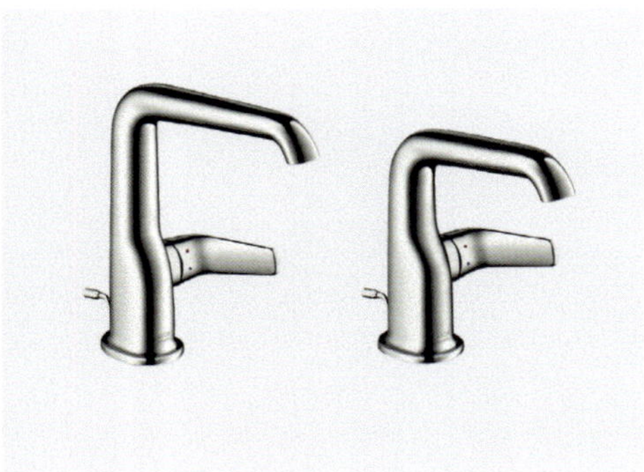

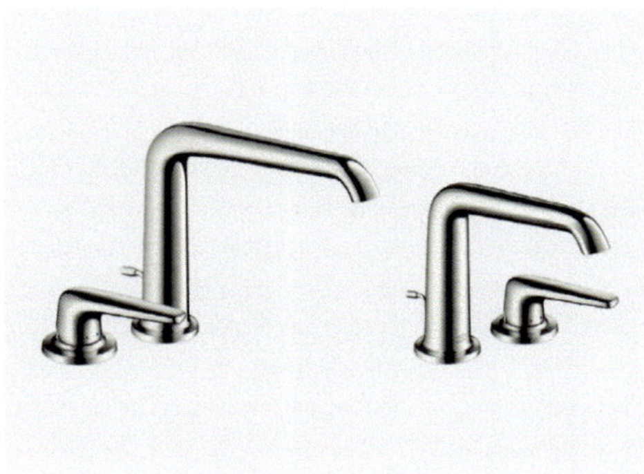

DA 7000 D Aura
Dunstabzugshaube

Extractor hood

Jury
Judges panel

Küche, Haushalt
Silber

Kitchen, household
Silver

Momentan gibt es kaum innovative Dunstabzugshauben auf dem Markt. Diese ist als reine Umluftvariante konzipiert. Sie funktioniert sehr gut, ist extrem leise und einfach zu bedienen. Sie besitzt Aktivkohlefilter und Fettfilter. Die Bedienelemente sind wunderbar minimalistisch integriert. Obwohl sie sehr groß ist, sieht sie ganz leicht aus und scheint zu schweben. Elegant und sehr ästhetisch.

There are scarcely any innovative extractor hoods on the market. But this one makes use of a new function – an extractor hood that works without extracting air. It is extremely quiet, works very well, and is simple to operate. It has charcoal filters and grease filters. The operating controls have been integrated in a wonderfully minimalistic way. Although it is very big, it looks quite light, and appears to float. Elegant and very tasteful.

Hersteller
Manufacturer
Miele & Cie. KG
D-33332 Gütersloh

Design
Designer
Werksdesign
In-house design

Vertrieb
Distributor
Miele & Cie. KG
D-33332 Gütersloh

Aura verbindet außergewöhnliches Design mit energieeffizientem Umluftbetrieb. Die innovative Dunstabzugshaube aus schwarzem Glas und weißen Abdeckungen bringt eine neue Formensprache in die Küche. Auffällig ist ihre elliptische Geometrie, die für eine schlanke und moderne Optik sorgt: In der Mitte ist der Haubenkörper gerade mal 16 Zentimeter hoch und läuft zu den Seiten flach aus.

Durch eine neue Konstruktion konnte ein Design ohne Kamin realisiert werden. So scheint die leicht wirkende und elegante Inselhaube über dem Herd zu schweben. Tatsächlich hängt sie an vier fast unsichtbaren Seilen, die den freien Blick in die Küche ermöglichen.

Die Besonderheit ist, dass Aura als reine Umluftvariante konzipiert ist. Dabei wurden großflächige, hoch effiziente Aktivkohlefilter eingesetzt, die Gerüche besonders gut binden. Zur leichten Reinigung ist die Haube mit spülmaschinengeeigneten Edelstahl-Metallfettfiltern ausgestattet.

Die neue Umlufthaube ist vor allem für die Installation in Niedrigenergie- und Passivhäusern prädestiniert, weil durch Heizsysteme erwärmte oder durch Klimaanlagen gekühlte Raumluft im Gebäude bleibt im Gegensatz zu Abluftsystemen, bei denen die Raumluft nach außen geführt wird. Aber auch in Mietwohnungen werden Umlufthauben aufgrund ihres einfachen und schnellen Einbaus bevorzugt.

Dank der neuen Motorentechnik ist sie extrem leise. Für einfache Handhabung sorgt die in die schwarze Glasblende integrierte Touchbedienung.

Aura combines extraordinary design with energy-efficient recirculated-air operation. Made of black glass with a white canopy, this innovative extractor hood brings a new formal expression to the kitchen. One striking feature is its elliptic geometry, which gives it a slim, modern appearance. At its centre, the body of the hood is just 16 centimetres high, and tapers from there to the edges.

A new design means that a flue is not needed. As a result, the light and elegant cooker hood appears to float above the hob. In fact, it is held in place by four almost invisible cables, so that the view into the kitchen remains unobstructed.

What is special about Aura is that it has been designed as a purely recirculating hood. To achieve this, large-surface, highly efficient charcoal filters have been used that trap odours especially well. For easy cleaning, the hood has been fitted with dishwasher-safe, stainless steel grease filters.

The new recirculating hood is perfect above all for use in low-energy or »passive« houses, as the air warmed by heating systems or cooled by air-conditioning systems remains in the building, unlike in extractor systems in which air is vented. But in rented apartments as well, recirculating hoods are preferred because they are simple and quick to install.

Thanks to its new motor technology, it is extremely quiet. The touch controls integrated in the black glass panel make it easy to operate.

Miele

SieMatic SF 10
Innenausstattung für Schubkästen und Auszüge

Accessories for drawers and pull-out units

Jury
Judges panel

Küche, Haushalt
Silber

Kitchen, household
Silver

Eine ausgesprochen schöne und hochwertige Innenausstattung von Küchenschubladen und Auszügen. Die von Kahla hergestellten Porzellangefäße lassen sich ebenso offen benutzen wie die übrigen Behältnisse. Sie sind in sich schöne Gegenstände, die man auch ins Regal oder auf den Esstisch stellen kann und nicht in der Schublade verstecken muss. Ästhetisch, schlicht, hochwertige Materialien und sehr praktische Details wie beispielsweise die Folienhalter. Super.

An outstandingly elegant and high-quality set of accessories for kitchen drawers and pull-out units. The porcelain containers made by Kahla can be used just as openly as the other containers. They are beautiful objects in their own right, that you can put on a shelf or dining table, and don't have to hide away in a drawer. Tasteful, unostentatious, high-quality materials and very practical details such as the kitchen-roll holders. Super

Hersteller
Manufacturer
SieMatic Möbelwerke GmbH
& Co. KG
D-32584 Löhne

Design
Designer
Speziell®
D-63067 Offenbach

Vertrieb
Distributor
SieMatic Möbelwerke GmbH
& Co. KG
D-32584 Löhne

Dieses neue Ausstattungssystem für Schubkästen und Auszüge ist bis ins kleinste Detail durchdacht. Es bietet optisch wie funktional zahlreiche vorbildliche Lösungen mit hohem Differenzierungspotential: von der Hightech-Konstruktion über die Präzision der Verarbeitung bis zum modernen Materialmix. Hell lasierte Eiche, edles Porzellan und hochwertiges Aluminium bilden eine elegante, filigrane Anmutung.

SieMatic SF 10 überrascht mit vielen ungewöhnlichen und raffinierten Ideen. Beispielsweise lassen sich Vorratsgefäße immer wieder neu mit Bleistift beschriften, sorgen GripDeck-Einlegeplatten mit Antirutschstreifen für sicheren Stand auch von schweren Gegenständen oder kann der Besteckeinsatz dank seiner ausklappbaren Tragegriffe leicht ins Esszimmer transportiert werden.

Stapelbare Holzboxen und vielfältig einsetzbare hochwertige Porzellangefäße in drei unterschiedlichen Höhen nutzen den Stauraum optimal. Die dekorativen Porzellan-Gewürzdosen mit verglastem Deckel lassen sofort den Inhalt erkennen und sehen zudem schön auf dem Esstisch aus. Daneben gibt es Halter für Alu- und Frischhaltefolien, ihre dezenten Porzellangriffe erleichtern das Öffnen. In der schräg angeordneten Porzellan-Messerbank können verschieden große Messer verstaut werden. Alle Porzellanelemente wurden gemeinsam mit dem deutschen Hersteller Kahla entwickelt.

Die vielfältigen Elemente lassen sich perfekt und individuell wie bei einem Baukasten zusammenfügen, sodass jeder seine ganz persönliche Ordnung schaffen kann.

This new system of accessories for drawers and pull-out units has been thought through down to the last detail. Visually and functionally, it offers exceptional solutions that set it apart from the crowd: from its high-tech design, to precise workmanship, to a modern mix of materials. Light-glazed oak, fine porcelain and high-quality aluminium create an elegant, delicate impression.

SieMatic SF 10 has many unusual and sophisticated ideas in store. For example, storage containers can be labelled and re-labelled using a pencil, GripDeck inserts with anti-slip strips ensure that even the heaviest of objects stand firm, and the cutlery insert with its fold-up handles can easily be carried into the dining room.

Stackable wooden boxes and versatile, high-quality porcelain containers in three different heights ensure that space is used in the best possible way. The decorative porcelain spice jars with their glass lids easily reveal their contents, and look good on the dining table as well. There are also holders for silver foil and cling film, whose unassuming porcelain handles make them easier to open. Knives of different sizes can be stored on the angled porcelain knife rack. All the porcelain elements were developed in collaboration with the German manufacturer Kahla.

The many elements can be combined perfectly and individually, like a kit, allowing everyone to create their own order.

SieMatic
by
KAHLA
SieMatic
welcome!

Depot
Gewürzset

Seasoning set

Jury
Judges panel

Tisch- und Küchenkultur
Silber

Tableware and kitchenware
Silver

Sehr überzeugend ist an diesem Gewürzset der Streuer. Egal, womit es gefüllt ist, man kann es mit einem Handgriff stufenlos dosieren – von ganz klein bis groß. Besonders gut: Man kann die Streuer auseinander schrauben. So lassen sie sich in der Spülmaschine reinigen.

What we especially like about this set is the different inserts. Whatever is in them, it can be dispensed as needed, from very small to large. We also like the way the jars can be taken apart. This means they can be cleaned in the dishwasher.

Hersteller
Manufacturer
WMF AG
D-73312 Geislingen

Design
Designer
Metz und Kindler Produktdesign
D-64293 Darmstadt

Vertrieb
Distributor
WMF AG
D-73312 Geislingen

Salz, Curry, Paprika, Pfefferkörner, Lorbeerblätter oder auch Kakao und Puderzucker sind in den eleganten Gewürzdosen und -streuern des Sets Depot ordentlich und übersichtlich aufgehoben.

Bei den vielseitig einsetzbaren Streuern stehen drei Lochbilder zur Verfügung: Für gemahlene Gewürze eignet sich der Universalstreuer. Wacholderbeeren, Rosmarin und ähnlich grobe Kräuter und Körner finden im Grobstreuer ihren Aufbewahrungsort, der eigens für sie entwickelt wurde, während sich für Paprika, Zimt und Co. der Feinstreuer mit seinem feinen Edelstahlsieb anbietet.

Aber auch für die nichtzerkleinerten Gewürze, wie etwa Nelken und Muskatnuss, gibt es eine Lösung. Sie sind in der Gewürzdose ohne Streubild gut aufgehoben.

Das Gewürzset ist aus Cromargan® und Glas gefertigt. Die zurückhaltend gestalteten und formschönen Streuer passen mit ihrem modernen und zeitlosen Design sowie ihrer Materialwahl in jede Küche. So schafft Depot nicht nur Ordnung, sondern die Gewürze sind in den Glasbehältern auch gut sichtbar und hygienisch untergebracht.

The elegant jars and shakers in the Depot set offer a neat and tidy way of storing salt, curry powder, paprika, peppercorns, and bay leaves, as well as cocoa and icing sugar.

Three different inserts are available for the versatile shakers. The universal shaker is suitable for ground spices. The coarse shaker with its large holes was designed especially for juniper berries, rosemary and other similar coarse herbs and spices, while the fine shaker with its fine stainless steel mesh was developed for paprika, cinnamon, and the like.

But there is also a solution for whole spices such as cloves and nutmeg. They can be kept in the spice jar without a shaker insert.

The seasoning set is made of Cromargan® and glass. With their modern, timeless design, as well as the materials chosen, these unassumingly shaped and elegant jars are a complement to any kitchen. Depot not only keeps the kitchen tidy, but also means that the herbs and spices are easy to find and stored away hygienically.

Poesie et Table. Breakfast
Butterdose

Butter dish

Tisch- und Küchenkultur
Silber

Tableware and kitchenware
Silver

Jury
Judges panel

Die Idee ist sehr charmant. Zum Beispiel heißt es »Butter bei die Fische«. Wenn die Butter zur Neige geht, sieht man die Fische. Dieses Produkt hat Humor: Wenn die Butter verbraucht ist und man sich eigentlich ärgern könnte, fängt man an zu schmunzeln. Schönes Detail: die Oberfläche – außen matt und innen glänzend.

The idea is very charming. So you have the saying on the lid, and when you get to the bottom of the dish, you find the image to go with it. Instead of being annoyed that the butter has all gone, you find yourself smiling. We like the way it is matt on the outside and glossy on the inside.

Hersteller
Manufacturer
Hartmut Räder
Wohnzubehör GmbH & Co. KG
D-44791 Bochum

Design
Designer
Werksdesign
In-house design
Tobias Langner, Sarah Baumann

Vertrieb
Distributor
Hartmut Räder
Wohnzubehör GmbH & Co. KG
D-44791 Bochum

Diese außergewöhnliche Butterdose bringt ein bisschen Poesie auf den Frühstückstisch. Auf dem Deckel steht in eingelassenem Schriftzug etwa »Butter bei die Fische«, »Butterblume« oder »Alles in Butter«.

Geht die Butter zur Neige, entdeckt man nach und nach den passenden Dekor auf dem Boden: rote Fische, gezeichnete Butterblumen oder viele schön arrangierte Gegenstände. Bei einer vierten Variante ist auf dem Deckel eine Kuh abgebildet, und auf der Unterseite wiederholen sich die Worte »moo. Muh«. Auf alle Fälle gibt es viel zu entdecken – kleine Botschaften mit viel Fantasie, die bei so manchem Morgenmuffel ein Lächeln hervorlocken.

Die formal schlichte und funktionale Butterdose besteht aus feinem Porzellan. Die klare Formsprache lässt die Ästhetik der unglasierten, samtigen Oberfläche mit dem eingeprägten Dekor voll zur Geltung kommen. Ein Produkt, das modernes, zeitloses Design mit spielerischen Komponenten verbindet.

This unusual butter dish brings a little poetry to the breakfast table. Embedded in the lid of three variants, there are various German sayings associated with butter.

And when the butter starts to run out and the dish has to be scraped, the images to go with these sayings appear on the base. And in a fourth variant, a cow is displayed on the lid, while the word »moo« is written on the base. At all events, there is a lot to discover – little messages with a lot of imagination, which will tease a smile from many a tired face first thing in the morning.

The formally plain and functional butter dish is made of fine porcelain. Its clear formal expression is the perfect foil to the aesthetics of the unglazed smooth surface with its embedded decoration. A product that combines modern, timeless design with playful components.

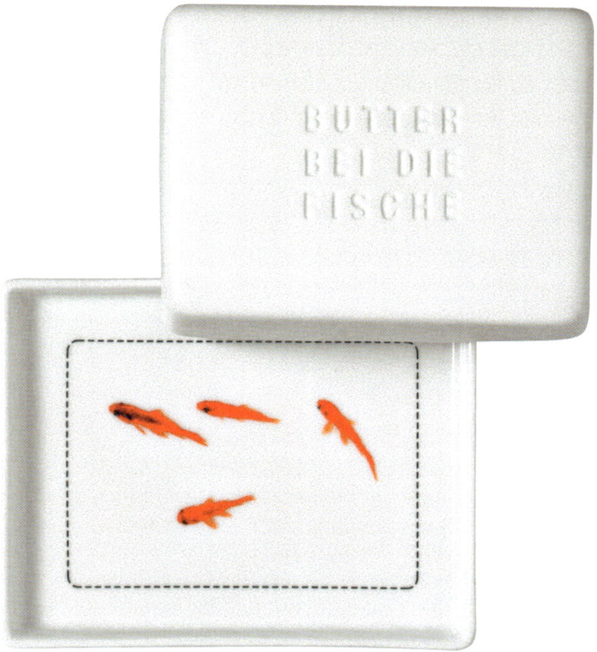
BUTTER
BEI DIE
FISCHE

moo. muh. moo. muh. moo. muh. moo. muh

Eva Solo Serie XO
Reiben

Graters and slicers

Jury
Judges panel

Tisch- und Küchenkultur
Silber

Tableware and kitchenware
Silver

Endlich mal eine Reibe, die einzeln benutzbar ist, trotzdem aufgestellt werden kann und nicht frei in der Luft gehalten werden muss. Sie ist stapelfähig und dadurch platzsparend verstaubar. Spülmaschinenfest. Einfache Form, zwei Materialien, nur ein Spritzgussteil und das Funktionsteil zugefügt. Sehr funktional und formschön.

At last a grater that can be used singly and with its feet firmly on the worktop, instead of in the air. It is stackable, and so saves space when stowed away. Dishwasher-safe. Simple form, two materials, just an injection-moulded part and the functional element on top. Very functional and tasteful.

Hersteller
Manufacturer
Eva Solo A/S
DK-2760 Måløv

Design
Designer
Tools Design
Henrik Holbæk, Claus Jensen
DK-2400 Kopenhagen NV

Vertrieb
Distributor
Eva Solo A/S
DK-2760 Måløv

Die Reiben in verschiedenen Varianten gehören zu der Eva Solo XO Collection, die sich an diejenigen wendet, die guten Geschmack mit den Augen und auf der Zunge erleben wollen. Es ist eine Serie von Küchengeschirr – Töpfe, Pfannen, Küchenutensilien und Zubehör –, die robuste Kochqualität mit einem markanten und stilreinen Design kombiniert, eine Zierde in jeder Küche.

Die Reiben sind in vier verschiedenen Varianten erhältlich: grob, medium und fein sowie als Hobel. Ob Apfel und Karotte, Gurke und Kartoffel oder Knoblauch und Parmesan – die extra scharfen, feinen Schneiden aus Edelstahl erfüllen die unterschiedlichsten Reibe-, Raspel- und Hobelaufgaben schnell und leicht.

Dank ihres exzellenten, ergonomischen Designs stehen die stabilen Reiben gut auf dem Tisch und liegen optimal in der Hand. Braucht man sie nicht, können sie einfach übereinander gestapelt und so platzsparend im Küchenschrank oder in der Schublade verstaut werden.

Obwohl die Reiben sehr robust und äußerst praktisch sind, wirken sie durch ihre hochwertige moderne Gestaltung leicht, elegant und formschön.

The various types of grater / slicers are part of the Eva Solo XO collection, which is designed for users whose good taste is both visual and gastronomic. It is a series of kitchenware – pots, pans, kitchen utensils and accessories – that combines sturdy, professional cook quality with a striking, uncluttered design. They are an adornment for any kitchen.

The graters / slicers are available in four different versions: there are coarse, medium and fine graters, and a slicer. Whether apple and carrot, cucumber and potato, or garlic and Parmesan – the extra sharp, fine stainless steel blades perform the widest variety of grating and slicing tasks quickly and easily.

Thanks to their excellent ergonomic design, the sturdy grater / slicers stand firmly on the worktop and fit perfectly into the hand. If they are not being used, they can simply be stacked on top of each other, and stowed away in the kitchen cupboard or drawer without taking up much space.

Although the grater / slicers are very robust and extremely practical, their high-quality, modern design means that the impression they give is one of lightness and elegance.

082
083

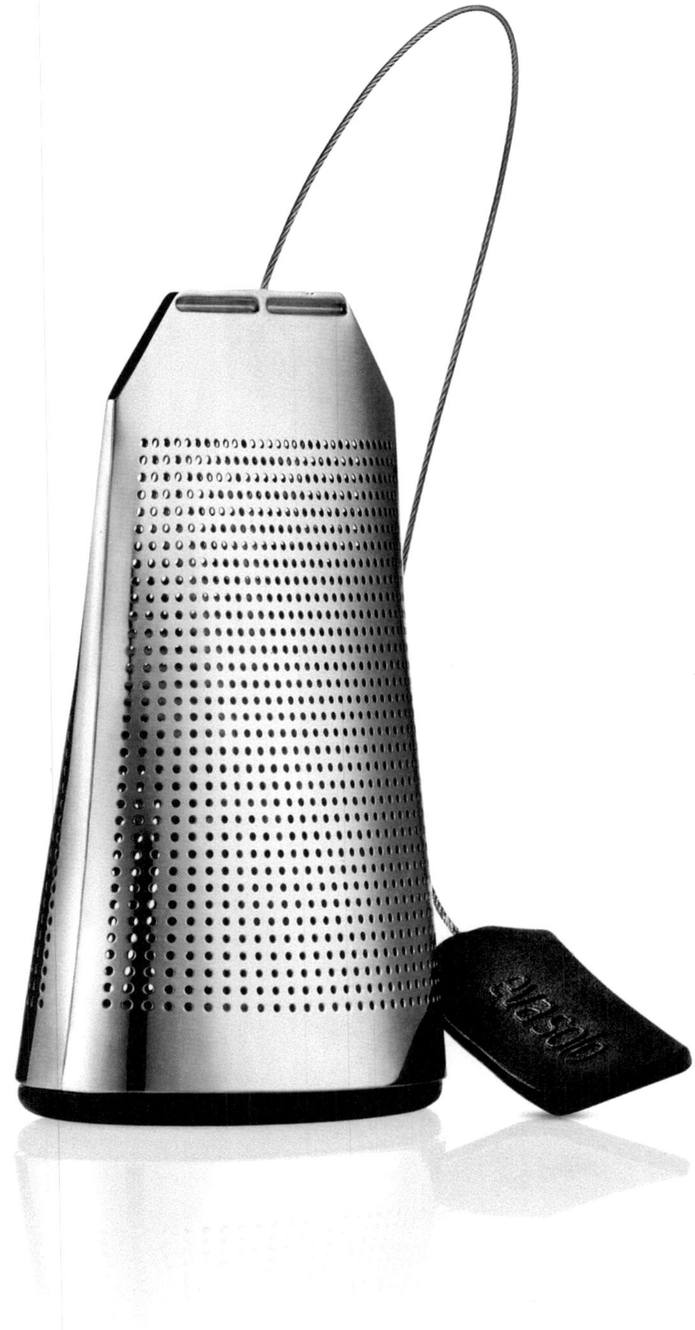

Eva Solo
Teebeutel

Tea bag

Jury
Judges panel

Tisch- und Küchenkultur
Silber

Tableware and kitchenware
Silver

Dieses Produkt greift die Formensprache eines Teebeutels auf und knüpft an die Erwartungshaltung der Nutzer an. Man kann es wieder befüllen, es ist ein Mehrweg- und kein Einwegprodukt. Als Ersatz für ein Tee-Ei finden wir das sehr geglückt. Eine charmante Spielerei.

It takes up the formal expression of a tea bag, and thus the expectations of the user. It can be refilled – a reusable product instead of a disposable one. It's a very good substitute for the conventional tea infuser. A charming gadget.

Hersteller
Manufacturer
Eva Solo A/S
DK-2760 Måløv

Design
Designer
Tools Design
Henrik Holbæk, Claus Jensen
DK-2400 Kopenhagen NV

Vertrieb
Distributor
Eva Solo A/S
DK-2760 Måløv

Der originelle Teebeutel greift die Form des klassischen Teebeutels auf, mit dem großen Unterschied, dass er im Gegensatz zu seinem Papier-Vorbild unendlich oft wieder aufgefüllt werden kann, denn er ist aus hochwertigem Edelstahl hergestellt. Der Deckel an der Unterseite besteht aus Silikon.

Er ist so konzipiert, dass das Teearoma von Schwarz-, Grün- oder Kräutertee optimal frei wird. In den Teebeutel können bis zu zehn Gramm Teeblätter eingefüllt werden – genug, um einen Liter Tee zu machen. Damit eignet er sich nicht nur für Tassen, sondern auch für Teekannen.

Zum Füllen des Teebeutels wird einfach der schwarze Silikondeckel an der Unterseite abgenommen. Nach dem Aufbrühen und Ziehen lässt er sich leicht an seiner flexiblen, langen Edelstahlschnur aus Tasse oder Teekanne wieder herausnehmen. Hinterher kann er in der Spülmaschine gereinigt werden.

Der ungewöhnliche Teebeutel ist aber nicht nur äußerst funktional und extrem robust, sondern auch formschön und ästhetisch gestaltet.

This original tea bag takes up the form of the classic tea bag, but with the big difference that, unlike its paper model, it can be refilled over and over again, since it is made of high-quality stainless steel. The lid on the underside is made of silicone.

It is designed so that the aroma of black, green and herb tea is released in the best possible way. Up to ten grams of tea leaves can be filled into the tea bag – enough to make one litre of tea. This makes is ideal not only for cups, but also for pots of tea.

All that has to be done to fill the tea bag is remove the silicone lid from the underside. After brewing, the tea bag's long, flexible stainless steel cord makes it easy to retrieve from the cup or pot. Afterwards, it can be cleaned in the dishwasher.

This unusual tea bag is not only extremely functional and robust, but also elegantly and tastefully designed.

Asia Line
Wok-Küchenhelfer

Wok kitchen aids

Jury
Judges panel

Tisch- und Küchenkultur
Silber

Tableware and kitchenware
Silver

Griffhaptisch gut gelöste Küchenhelfer, die über ihre Grundgeometrien überraschende, praktische Funktionen in der Handhabung besitzen. Sie beziehen sich auf die asiatische Kochkultur, und das spiegelt sich besonders in dem Löffel wieder. Er wirkt wie Origami. Schöne Kunststoffspritzelemente. Gutes Preis-Leistungs-Verhältnis.

Kitchen aids that are good to grip and pleasant to the touch. Their basic geometries give them surprising and practical functions when in use. They allude to Asian culinary culture, and this is especially reflected in the strainer. It almost appears to be folded, like origami. Appealing injection-moulded plastic elements. Good value for money.

Hersteller
Manufacturer
Fackelmann GmbH & Co. KG
D-91217 Hersbruck

Design
Designer
Werksdesign
In-house design
Fabian Graf von Reichenbach

Vertrieb
Distributor
Fackelmann GmbH & Co. KG
D-91217 Hersbruck

Die asiatische Küche hat schon viele begeisterte Liebhaber gefunden. Asia Line ist ein Küchenzubehör, das das Kochen asiatischer Gerichte unterstützt. Ohne Kratzspuren in beschichteten Woks zu hinterlassen, lässt sich mit den praktischen und formschönen Küchenhelfern Gemüse, Nudeln, Soßen, Fleisch, Fisch und vieles andere zubereiten. Die Wok-Serie besteht aus Nudel-, Rühr- und Schöpflöffel sowie Abseihschaufel und Schlitzwender.

Unter diesen Küchenwerkzeugen sticht besonders die multifunktionale Abseihschaufel hervor. Sie eignet sich zum Kochen, Abseihen und Servieren. Da der Rand der Schaufel hochgezogen ist, kann sie zum »Pfannenrühren« verwendet werden – eine typische Tätigkeit beim Kochen im Wok. Durch das Schneeschaufelprinzip ist so ein schnelles Umschichten der Speisen möglich. Und dank ihrer Perforierung, die die Form von Reiskörnern hat, lassen sich mit der Abseihschaufel auch Lebensmittel, die in Brühe oder Sauce zubereitet werden, bequem abschöpfen.

Die Wok-Küchenhelfer werden alle in Deutschland hergestellt und bestehen aus stabilem Nylon, das bis zu 200 Grad hitzebeständig ist.

Asian cuisine has found many enthusiastic adherents. Asia Line is a series of kitchen accessories that help cook Asian dishes. These practical and elegant kitchen aids help cooks prepare vegetables, noodles, sauces, meat, fish, and much more without scratching the wok's non-stick surface. The wok series comprises spoons and ladles as well as a straining scoop and slotted turner.

Of these kitchen implements, the multifunctional straining scoop stands out particularly. It can be used to cook, strain and serve. As the scoop's edges are raised, it can be used for stir frying, a typical method when cooking with a wok. This »snow shovel« technique allows the food in the wok to be quickly turned over. Thanks to its perforation, with holes shaped like grains of rice, the straining scoop can be used to lift out food prepared in broth or sauce.

All the wok kitchen aids are made in Germany, and are made of robust nylon that is heat-resistant up to 200 degrees Celsius.

Flachschmieder
Universal Küchenmesser

Universal kitchen knife

Jury
Judges panel

Tisch- und Küchenkultur
Silber

Tableware and kitchenware
Silver

Ein Messer, das von seiner Wertanmutung her überzeugt und ergonomisch gut in der Hand liegt. Geeignet für Links- wie Rechtshänder. Ein archaisches Produkt – so sieht ein Messer aus. Man erkennt die Handarbeit. Das schöne Detail, wie die kleine Stufe am Übergang von Klinge und Griff, ist ein spezielles Merkmal dieses Messers.

A knife with an impressive high-quality appearance. Ergonomically, it sits well in the hand. Left and right-handers can work with it. An archaic product – this is what a knife should look like. You can see that it is hand-made. The nice detail of a small step at the transition between blade and handle is a distinguishing feature of this knife.

Hersteller
Manufacturer
Windmühlenmesser
Robert Herder GmbH & Co. KG
D-42697 Solingen

Design
Designer
Werksdesign
In-house design
Giselheid Herder-Scholz

Vertrieb
Distributor
Windmühlenmesser
Robert Herder GmbH & Co. KG
D-42697 Solingen

Dieses Messer ist ein Botschafter sowohl für das Messerhandwerk als auch für das Flachschmieden, das bereits zu den verschwindenden Gewerken gehört. Giselheid Herder-Scholz hat in Zusammenarbeit mit Solingens letztem Flachschmieder Rudolf Broch ein besonderes Messer gefertigt.

So wurde die Messerform speziell für den Hammer des Handwerkers entworfen. Er hat die Klinge von Hand geschmiedet und beschnitten, gelocht und geglüht. Nach dem Härten wurde es bei Robert Herder von Hand walkgeschliffen und blaugepließtet – die aufwendigste Form traditionellen Solinger Feinschleifens. Dadurch wird die Klinge besonders glatt und bekommt einen bläulichen Regenbogenschimmer, weil feinste Schleifspuren das Licht brechen. Anschließend wird sie mit Griffschalen aus Walnußwurzelholz versehen, danach fein geglättet und geölt.

Der Vorteil einer Klinge, die mit vielen Schlägen von Hand sorgfältig ausgeschmiedet wurde, ist das dichtere Gefüge, das man vor allem im Schneidebereich erreichen kann. Das garantiert eine hohe Verschleißfestigkeit und eine langanhaltende Schärfe.

Am Übergang von Klinge und Griff befindet sich eine kleine Stufe, der Einsatz, das typische Zeichen flachgeschmiedeter Messer. Bewusst wurde hier ein Stück der Klinge mit schwarzem Zunder belassen.

Das Messer aus hoch kohlenstoffhaltigem Stahl vereint in sich sowohl die Charakteristika dieser besonderen, alten Schmiedeart als auch der traditionellen Schleiftechniken der Firma Robert Herder – beides Kenntnisse und Fertigkeiten, die es wert sind, erhalten zu werden. Ein Paradebeispiel Solinger Schleif- und Schmiedekunst.

This knife is an ambassador both for the knife-making craft and for flat-forging, which is already regarded as a dying art. In collaboration with Rudolf Broch, Solingen's last flat-forger, Giselheid Herder-Scholz has come up with a special knife.

The form of the blade was designed especially for the craftsman's hammer. By hand, he forged, cut, punched and annealed the blade. Following hardening, it was ground and honed by hand at Robert Herder – the most complex process of traditional finish-grinding in Solingen. As a result of this process, the blade becomes especially smooth and takes on a bluish rainbow sheen, due to the light being refracted by the extremely fine sanding scratches. Following this, the blade is fitted with a handle of walnut root wood, then finely smoothed and oiled.

The advantage of a blade that is carefully forged by hand is the denser structure that can be achieved, especially along the cutting edge. This ensures extreme wear-resistance and long-lasting sharpness.

At the transition between the blade and the handle there is a small step – the typical sign of a flat-forged knife. Here, part of the blade has deliberately been left with a black oxidation layer.

Made of steel with a high carbon content, the knife brings together the typical details of this special, ancient forging craft and the traditional honing techniques of Robert Herder – both abilities and skills that are worth preserving. A prime example of Solingen's honing and forging art.

HRC 60
Solingen

Kamo-to Edition Lignum S2 Kochmesser

Chef's knife

Jury
Judges panel

Tisch- und Küchenkultur Silber

Tableware and kitchenware Silver

Das hochwertige Messer besteht aus zwei Materialien und ist vollständig in Handarbeit gefertigt. Mit Signatur des japanischen Schmiedes. Ein Klassiker. Gelungene Verbindung von japanischer Schmiedekunst und Solinger Griffbearbeitung.

This high-quality knife comprises two materials, and is made completely by hand. With the signature of the Japanese cutler. A classic. Well executed combination of Japanese forging art and German handle-making.

Hersteller
Manufacturer
Windmühlenmesser
Robert Herder GmbH & Co. KG
D-42697 Solingen

Design
Designer
Werksdesign
In-house design
Tomoyuki Takada, Shirou Kamo

Vertrieb
Distributor
Windmühlenmesser
Robert Herder GmbH & Co. KG
D-42697 Solingen

Das Lignum S2 gehört zu dem Konzept »Messerhandwerk Ost und West«. Ende der 1990er Jahre knüpfte die Firma Robert Herder die ersten Kontakte mit japanischen Messerschmieden. Daraus ergab sich eine außergewöhnliche und langjährige Zusammenarbeit, die unterschiedlichste Impulse gab und gibt. Seither hat viel gegenseitiger Austausch untereinander stattgefunden, sodass es nur eine Frage der Zeit war, dass die Erfahrung und das Können von beiden Seiten in einem ganz neuen gemeinsamen Projekt Gestalt annahm – und zwar in einem Kooperationsmesser.

Dieses Messer ist das Lignum, dessen Klinge von dem Schmied Shirou Kamo in Japan geschmiedet, gehärtet und geschliffen wird. Die Klinge ist ein Schmiedeverbund aus zwei verschiedenen rostfreien Stählen. Der Klingenkern besteht aus dem extrem harten rostfreien Stahl VG 5 und der Außenmantel aus rostfreiem 410er-Stahl.

Während die Klinge in Japan bearbeitet wird, kommt das Produktdesign und die Griffbearbeitung des mittelgroßen Kochmessers aus Solingen. Die Griffschalen werden dann sorgfältig in Handarbeit genau angepasst und ausgeformt. Sie bestehen aus edlen Hölzern: wahlweise aus Makassar-Ebenholz, Nussbaum, Amboina und Amazonas-Palisander.

Das kompakte, etwas breitere Kochmesser ist ein Vielzweckmesser zum Schneiden von Obst oder Gemüse, Filetieren von Fisch oder Zerteilen von Fleisch.

The Lignum S2 is part of the »Knife-making craft in the east and west« concept. At the end of the 1990s, Robert Herder established its first contacts with Japanese cutlers. This gave rise to an unusual and long-lasting collaboration that has given and continues to provide diverse stimuli. Since then there has been much mutual exchange of experience and information, and so it was only a matter of time before the experience and skills of the two parties took on the form of a completely new project – to jointly produce a knife.

The knife in question is the Lignum, whose blade is forged, tempered and honed in Japan by the cutler Shirou Kamo. The blade is a composite of two different stainless steels. The blade core is made of extremely hard VG 5 stainless steel, and the outer sheath of 410 stainless steel.

While the blade is processed in Japan, the handle of this medium-sized chef's knife is made in Solingen, where the product is also designed. The two parts of the handle are then carefully and precisely fitted and shaped by hand. The handle is made of precious wood, such as Makassar ebony, walnut, amboina and Amazon rosewood.

The compact, somewhat broad chef's knife is a multipurpose implement for chopping fruit or vegetables, filleting fish or cutting meat.

Jury
Judges panel

Knuff
Baby- und Kinderbesteck

Baby's and child's cutlery

Tisch- und Küchenkultur
Silber

Tableware and kitchenware
Silver

Was uns an dem Kinderbesteck besonders gut gefällt, ist, dass es nicht mit den für Kinderbestecke üblichen verniedlichten Dekoren arbeitet. Es ist klassisch und sieht auf dem Tisch der Erwachsenen gleichberechtigt aus, nur ein wenig kleiner und mit etwas veränderten Proportionen. Die abgerundete, schmälere Vorderseite macht es angenehm zum Essen, weil Kinder eher von vorne und nicht von der Seite essen. Schöne Anmutung durch den matten Edelstahl, fühlt sich weicher an, auch vom Gewicht her.

What we especially like about this children's cutlery is that is doesn't use the usual twee decoration you find with such cutlery sets. It is classic in appearance, and seems to take on equal status at the grown-ups' table. It is simply a little smaller, with somewhat different proportions. The rounded, narrower front element makes eating pleasant, because children tend to eat from the front rather than from the side. Graceful in appearance thanks to the matt stainless steel. It feels softer, also because of its weight.

Hersteller
Manufacturer
Auerhahn Bestecke GmbH
D-72213 Altensteig

Design
Designer
Kurz Kurz Design
D-42697 Solingen

Vertrieb
Distributor
Auerhahn Bestecke GmbH
D-72213 Altensteig

Knuff ist ein Esslernbesteck, das vom ersten Breilöffel zum Füttern des Babys bis zu den selbstständigen Essversuchen des Kindes eingesetzt werden kann. Die Produktfamilie reicht vom Babylöffel über Speiseschieber und Querlöffel bis zum vierteiligen Essbesteck.

Dabei sind alle Teile ergonomisch ausbalanciert. Die kompakten Griffe schmiegen sich perfekt in Kinderhände. Die Vorderteile wurden bewusst schauflig und mit runden Ecken gestaltet.

Das Besteck bietet auch Multifunktionsvorteile: So hat das Messer eine Klingenform, mit der sich sowohl kindgerecht schieben als auch schneiden lässt. Und die Gabel kann dank ihrer Mulde gleichzeitig zum Löffeln verwendet werden.

Das spülmaschinenfeste Baby- und Kinderbesteck besticht auch durch seine zeitlose und moderne Gestaltung ohne Dekor. Nicht nur wegen seines zurückhaltenden und ästhetischen Designs, sondern auch dank seiner hohen Material- und Verarbeitungsqualität – es besteht aus mattem Edelstahl – passt es von Anfang an in die gehobene Welt der Tischkultur und zu den Bestecken der Erwachsenen. Und die kleinen Leute können so mit einem Besteck essen, das dem der großen Leute sehr nahe kommt – und sich fast wie die Großen fühlen.

Knuff is cutlery that teaches the child how to eat, and can be used from the first time the baby is fed porridge to the child's first attempts to eat on its own. The product family ranges from a baby's spoon to a pusher and offset spoon, to a four-piece cutlery set.

All the elements are ergonomically balanced. The compact handles fit perfectly into a child's hand. The front parts have deliberately been given a shovel-like form and rounded corners.

The cutlery has the advantage of being multifunctional. The shape of the knife blade is such that children can use it to push or cut food. And thanks to its hollow form, the fork can also be used as a spoon.

Dishwasher-safe, the baby's and child's cutlery is striking for its timeless and modern design, without any decoration. Thanks not only its unassuming and tasteful design, but also to the high quality of the materials and workmanship – it is made of matt stainless steel – it harmonizes right from the start with the sophisticated tableware and cutlery of the adult world. In this way, the very young can eat using cutlery that is very similar to that of the adults – and feel almost grown up.

Setdown4
Topfrost

Trivet

Tisch- und Küchenkultur
Silber

Tableware and kitchenware
Silver

Jury
Judges panel

Eine klassische Struktur eines Topfrostes in neuem Material, Edelstahl, sehr hochwertig verarbeitet. Er besitzt eine schöne Anmutung, die an eine Fahrradkette erinnert. Ausgesprochen attraktives und überzeugendes Produkt. Auch die Verpackung ist sehr hochwertig.

A classic trivet structure in a new material, stainless steel: excellent workmanship. It has an appealing air, reminiscent of a bicycle chain. Outstandingly attractive and impressive product. Its packaging is also of very high quality.

Hersteller
Manufacturer
Cap. GmbH
D-32609 Hüllhorst

Design
Designer
Werksdesign
In-house design
Freimut Stehling

Vertrieb
Distributor
Cap. GmbH
D-32609 Hüllhorst

Der Topfrost Setdown4 erinnert an ein fein gearbeitetes Schmuckstück. Elegant, filigran, technisch anmutend, das sind nur ein paar der Charakteristika des außergewöhnlichen Topfrostes. Er besteht aus schlanken Rundstangen, die miteinander verbunden sind.

Auf ihm lassen sich direkt neben dem Herd in unmittelbarer Nähe zum Kochen die heiße Pfanne, die große Platte oder der schwere Bräter abstellen – denn Setdown4 misst 47 mal 30 Zentimeter. Den Topfrost kann man aber auch als attraktive Ablage für den Esstisch verwenden.

Er besticht durch ein weiteres funktionales Produktmerkmal: Braucht man ihn nicht, lässt er sich einfach zusammenschieben oder rollen und beansprucht dann zur Aufbewahrung nur noch ganz wenig Platz. So klein kann er beispielsweise in einer Schublade untergebracht werden.

Durch sein ästhetisches, reduziertes Design fügt er sich in jedes Umfeld ein. So lässt er sich neben der Kochstelle, auf der Theke oder dem Esstisch platzieren.

Der Topfrost ist hochwertig verarbeitet und besteht aus Edelstahl. Dank seines Materials ist Setdown4 lebenslang zu gebrauchen.

The Setdown4 trivet is reminiscent of a finely worked piece of jewellery. Elegant, delicate, with a technical air – these are just a few of the characteristics of this unusual trivet. It is made of slim rods connected to each other.

It can be placed immediately next to the cooker, and hot pans, large platters or heavy roasting pans can be set down on it – Setdown4 measures 47 by 30 centimetres. But this trivet can also be used as an attractive mat for the dining table.

There is a further functional characteristic that stands out. If it is not in use, the pieces can be pushed or rolled together, so that it takes up only very little space when stowed away. It can easily be kept in a drawer, for example.

Its tasteful, reduced design means that it harmonizes with any surroundings. It can be placed next to the hob, on a counter or on the dining table.

The trivet is the result of excellent workmanship, and is made of stainless steel. Thanks to this material, Setdown4 will last a lifetime.

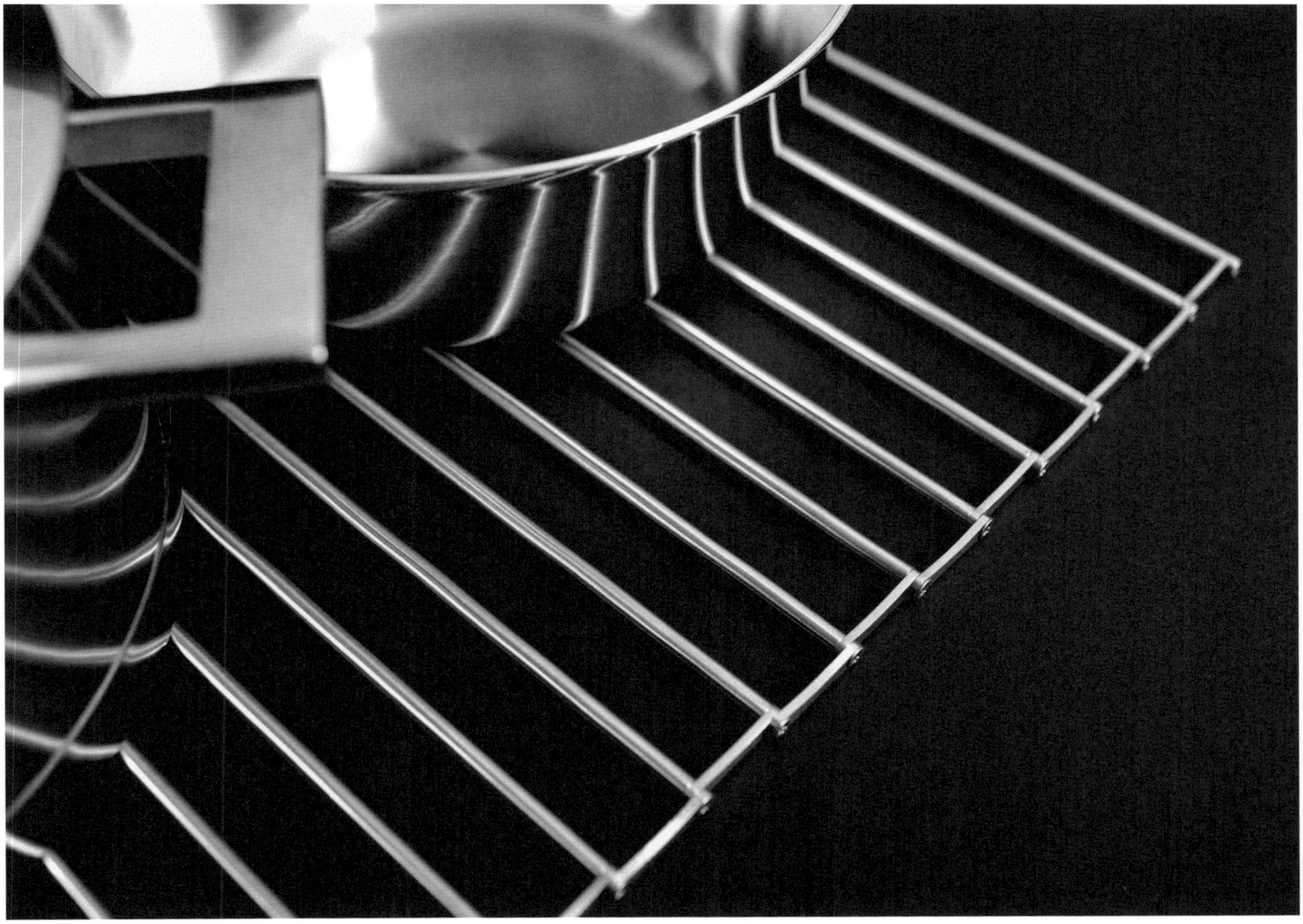

Eiswürfel-Max
Eiswürfelbereiter mit Deckel

Ice-cube maker with lid

Jury
Judges panel

Tisch- und Küchenkultur
Silber

Tableware and kitchenware
Silver

Der Eiswürfelbereiter zeigt einen wichtigen Aspekt. Er besitzt einen Klappdeckel, der verhindert, dass man Wasser verschüttet. Weiteres Plus: Das Eis nimmt nicht den Geruch des Kühlschranks an. Interessantes Konzept.

The ice-cube maker has an important new feature. It comes with a hinged lid, which means water does not get spilled. A further plus point is that the ice-cubes do not take on any odours from the fridge. Interesting idea.

Hersteller
Manufacturer
TRI International
D-88339 Bad Waldsee

Design
Designer
Gerhard Reichert
D-88719 Stetten

Vertrieb
Distributor
Versandhaus Walz GmbH
D-88339 Bad Waldsee

Wer kennt das nicht: Man möchte Eiswürfel machen, füllt den Eiswürfelbehälter mit Wasser und balanciert ihn zum Kühlschrank. Meist geht das nicht ohne lästige Wasserstraßen.

Dies passiert dagegen nicht beim Eiswürfel-Max. Dieser Eiswürfelbereiter besitzt zum einen einen Deckel, zum anderen ein bemerkenswertes wie praktisches Funktionsprinzip: einen in die Abdeckung integrierten pfiffigen Klappdeckel. Dieser lässt sich leicht öffnen und der Behälter kann bequem am Wasserhahn befüllt werden. Mit geschlossenem Deckel lässt sich der Eiswürfelbehälter sicher transportieren und ins Gefrierfach stellen, ohne dass Wasser verschüttet wird. Außerdem nehmen die Eiswürfel dadurch nicht den Geruch des Kühlschranks an.

Die zurückhaltend gestaltete Box mit ihren handlichen Tragegriffen und den abgerundeten Ecken ist auf ihre praktische Funktion reduziert. Der spülmaschinenfeste Eiswürfelbereiter aus Kunststoff bietet Platz für vierzehn Eiswürfel. Weiche Wölbungen und die konischen Formen des Behälters ermöglichen es, die Eiswürfel einfach und leicht zu entnehmen.

We all know the problem: you want to make ice-cubes, fill the ice-cube tray with water and then have to carry it over to the fridge. In most cases, water gets spilled along the way.

But not with Eiswürfel-Max. For one thing, the ice-cube maker has a lid. But it also has a functional principle that is as remarkable as it is practical: a clever hinged lid integrated into the cover. It is easy to open, and the container can conveniently be filled from the tap. With the lid closed, the ice-cube maker can safely be transported and placed in the freezer compartment without any water being spilled. In addition, this means the ice-cubes do not take on any odours from the fridge.

Unassuming in design, the box is reduced to its practical function, with convenient handles and rounded corners. Made of dishwasher-safe plastic, the ice-cube maker has room for 14 ice cubes. The slight camber and conical form of the vessel make it easy to take out the ice cubes.

RM
Raummodul

Room module

Jury
Judges panel

Objekt
Silber

Contract furniture
Silver

Ein ganz wichtiges Produkt, um große Räume offen zu nutzen. Mit dem Raumteiler lassen sich viele Funktionen abdecken und Räume gestalten. Es ist ein Wechselspiel aus Transparenz und Dichtheit. Die Struktur kann verschiedenste Elemente aufnehmen, zum Beispiel nur flächig, mit Schallschutz oder Regale, Tische et cetera. Wenn man Stauraum von beiden Seiten nutzen will, braucht man normalerweise eine große Tiefe. Hier ist das möglich mit der Hälfte der Tiefe. Es ist eine schöne Art zu teilen.

A very important product for using large offices in an open way. The partitioning wall covers many functions, and allows the room to be designed properly. It provides an interplay of transparency and opacity, and elements of all kinds can be added to the basic structure. It can either be a simple panel, or sound insulation, shelves, desks, etc. Normally if you want to use storage space from both sides, you need furniture with a lot of depth. Here, this is possible in just half the depth. It is a nice way of dividing up space.

Hersteller
Manufacturer
Bene AG
A-3340 Waidhofen / Ybbs

Design
Designer
Johannes Scherr
D-1070 Wien

Vertrieb
Distributor
Bene Stuttgart
D-70565 Stuttgart

Wer offene Bürolandschaften unterteilen will, braucht in der Regel Stellwände und dazu Regale zum Stauen. Das RM Raummodul bietet dagegen beides in einem und noch viel mehr. Es gliedert Büros, öffnet sich für Kooperation und Kommunikation, schafft Arbeitskomfort und Flächeneffizienz zugleich.

Herzstück ist ein stabiler Aluminiumrahmen, der sich auf beiden Seiten mit Paneelen, Boxen, Regalen, Schränken, Tischen, Glas- und Plexiglasbändern, Öffnungen, Leuchten und anderem Mobiliar bestücken lässt. Seine einfachste Anwendung ist eine Stellwand. Eingehängte Wandschränke erweitern sie um Staufläche, Glaselemente und Öffnungen bringen Durchblick. Lampe und Tisch machen RM zum Arbeitsplatz. Mit Wandschränken und Abdeckpaneelen können intime Kojen gebaut werden.

Regale bieten Akten, Pinnwände der Persönlichkeit Platz. RM Raummodule lassen sich additiv erweitern, an ihnen können mehrere Arbeitsplätze andocken. Vis-à-vis aufgestellt, bilden sie als Wandscheiben den Rahmen für Teamwork. Eine von ihnen ließe sich beispielsweise als Bibliothek befüllen und rückseitig mit einem Stehtisch bestücken: zum Entspannen und Plaudern.

Der flexible Raumteiler bietet nicht nur bei seinen Funktionen zahlreiche Variationen, sondern auch bei Farben und Materialien wie Melamin, Furnier, Glas, Plexiglas und bunte Stoffe sowie seinen Maßen.

As a rule, anyone who wants to divide up open-space offices needs partitioning walls and shelves for storage. The RM room module offers both in one, and much more besides. It gives offices a structure, opens up space for collaboration and communication, creates comfortable workplaces and helps use the available space efficiently.

The heart of the unit is its stable aluminium frame, which can be equipped on both sides with panels, boxes, shelves, desks, glass and Plexiglas panels, open spaces, lights, and other items of furniture. Its simplest use is as a partitioning wall. Adding cupboards to it creates storage space, while glass elements and openings create a clear field of vision. A lamp and a desk make RM into a workplace. Using cupboards and panels, intimate cubicles can be created.

Shelves offer space for files, while notice boards allow room for personal touches. RM room modules can be added to and extended, with several workplaces being docked onto them. If they are set up facing each other, the wall spaces they create provide space for teamwork in the middle. One of them can, for example, be filled with books to make a library, and fitted with a lectern on the other side: for relaxation and conversation.

And it is not only functionally that this flexible partitioning wall offers many variants. It also comes in different colours and materials, such as melamine, veneer, glass, Plexiglas, and bright textiles. Its dimensions vary as well.

Macao
Cafeteriastuhl

Cafeteria chair

Objekt
Silber

Contract furniture
Silver

Jury
Judges panel

Das ist ein Cafeteriastuhl, der extrem bequem ist. Er fordert einen zum Setzen auf. Uns gefällt, dass bei der Version mit Armlehnen diese eine einladende Geste machen, indem sie sich nach außen öffnen. Dadurch hat man das Gefühl, dass man nicht eingeengt wird. Es ist ein schlichter und funktionaler Stuhl. Leicht, stapelbar, exzellent gemacht.

This is an extremely comfortable cafeteria chair. It demands to be sat on. We like the way the version with the armrests invites you to come and sit, the way it opens up to the outside. As a result, it doesn't give you the feeling of being hemmed in. The chair is unostentatious and functional. Light, stackable, excellently done.

Hersteller
Manufacturer
Wiesner-Hager Möbel GmbH
A-4950 Altheim

Design
Designer
Neunzig° Design
D-73240 Wendlingen

Vertrieb
Distributor
Wiesner-Hager Möbel GmbH
D-97080 Würzburg

»Organische Gestaltung bedeutet, das Wesen der Natur zu untersuchen, ihre innere Logik zu verstehen und in unsere Produktwelt einfließen zu lassen. Die Natur arbeitet nicht präzise, aber sehr effektiv... Statt computergenerierter Perfektion entdecken wir die Schönheit frei modellierter Formen.« Die Natur war denn auch Vorbild bei der Gestaltung von Macao. So ließen sich die Designer von der eleganten Ästhetik und der hochfunktionalen Konstruktion natürlicher Strukturen inspirieren.

Herausgekommen ist ein leichter, filigraner Cafeteriastuhl, der konventionelle Formleitbilder durchbricht. Die Oberseite der Kunststoffschale besitzt eine klare fließende Kontur. Die Beine scheinen aus der Unterseite der Schale herauszuwachsen. Er zeichnet sich einerseits durch hohe gestalterische Qualität, ein zartes und leichtes Erscheinungsbild sowie ausgezeichneten Sitzkomfort aus, andererseits durch eine tragfähige robuste Konstruktion.

Von den Aufnahmepunkten zum Zentrum der Sitzfläche ist die Kunststoffschale verstärkt. Es entsteht ein stabilisierendes Kreuz, das die notwendige Tragfähigkeit garantiert. Die Beine aus lackiertem Präzisions-Stahlrohr sind einzeln mit der Sitzschale verschraubt. Der stapelbare Stuhl, den es mit und ohne Armlehnen gibt, kann auch mit unterschiedlichen Sitzpolstern bezogen werden.

Macao wird in allen Bereichen der Regeneration eingesetzt, neben Cafeterien auch in Restaurants, Bistros, Pausen- und Aufenthaltsräumen.

»Biomorphic design means examining how nature works, understanding its inner logic, and allowing this to inform our product world. Not only does nature work precisely, it is also very effective. Instead of computer-generated perfection, we discover the beauty of freely modelled forms.« Nature was also the model when designing Macao. The designers took their inspiration from the elegant aesthetics and highly functional composition of natural structures.

The end result is a light, delicate cafeteria chair that breaks with conventional formal principles. The upper side of the PVC shell has clearly flowing contours. The legs appear to grow out of the underside of this shell. The chair stands out by virtue of its high-quality design, it delicate and light appearance, and its outstanding comfort, but not only because of this. Its design is also sturdy and strong.

The shell is reinforced from the corners to the centre of the seat, creating a stabilizing support that gives it its load-bearing strength. The legs, which are fashioned out of lacquered precision steel tubing, are each bolted individually to the seat shell. This stackable chair, which is available with and without armrests, can also be upholstered with different seat cushions.

Macao can be used in recreational areas of all kinds – not only in cafeterias, but also in restaurants, bistros, common rooms and lounges.

Hersteller
Manufacturer
Olbrish Produkt GmbH
D-10961 Berlin

Design
Designer
Werksdesign
In-house design
Wolfgang Olbrish

Vertrieb
Distributor
Olbrish Produkt GmbH
D-10961 Berlin

WAVE OLBRISH PRODUKT

Wave
Handtaschenserie

Ambiente, Lifestyle
Gold

Handbag series

Ambience, lifestyle
Gold

Wave
Handtaschenserie

Handbag series

Ambiente, Lifestyle
Gold

Ambience, lifestyle
Gold

Jury
Judges panel

Formschön, zeitlos und langlebig. Durch die Geometrie der Außenhülle ergibt sich die Möglichkeit, nur einen Haltegriff zu verwenden, weil er über die Diagonale geht. Wir finden das hervorragend. Das ist etwas Innovatives und funktional Begründbares in dem ansonsten relativ beliebigen Handtaschensektor. Endlich mal eine Handtasche, bei der die zweite Schlaufe nicht ständig von der Schulter rutscht.

Elegant, timeless, and long-lasting. The geometry of the outer skin allows just one handle to be used, as it is placed diagonally. A super idea. An innovative, functionally sensible bag in a product area in which, generally speaking, anything goes. At last, a handbag where the second strap does not constantly slip off your shoulder.

Hersteller
Manufacturer
Olbrish Produkt GmbH
D-10961 Berlin

Design
Designer
Werksdesign
In-house design
Wolfgang Olbrish

Vertrieb
Distributor
Olbrish Produkt GmbH
D-10961 Berlin

Nomen est omen. So basiert Wave auf der Form einer sich brechenden Welle, die der Boden und dadurch auch die Wände der Tasche übernehmen. Der Reißverschluss greift das zentrale Thema in seinem Verlauf von der Vorder- zur Rückwand ebenfalls auf.

Die Tasche braucht keine Seitenteile dank der Form ihres Bodens. Dadurch wirkt sie schmal und filigran, ohne dass das benutzbare Taschenvolumen reduziert wird. Zusätzlich zum Hauptfach mit Reißverschluss bietet im Inneren ein Fach Platz fürs Handy. Die beiden größeren Taschen besitzen außerdem ein großes Reißerfach.

Die Zweifarbigkeit akzentuiert die gegenläufige Bewegung der Rück- und Vorderwand und greift in den Hell-Dunkel-Farbansätzen das Licht- und Schattenspiel einer Welle auf. Durch die Verdrehung der Seitenwände und das Schmalerwerden nach oben bietet Wave einen hohen Tragekomfort.

Die Tasche, die in Berlin umweltgerecht produziert wird, ist für eine mindestens zehnjährige Lebensdauer entwickelt worden. Ihre Außenhaut besteht aus hochwertigem süddeutschen Rindsnappaleder, das durchgefärbt und farbstabil, lichtecht und wasserabweisend ist. Innen ist Wave mit Veloursleder gefüttert, das ebenfalls Langlebigkeit gewährleistet.

Die Taschen, die es in drei verschiedenen Größen gibt, sind vielseitig einsetzbar: die kleinste zum Ausgehen, die mittlere für den täglichen Gebrauch und die große für Dokumente, Notebook und den kleinen Einkauf.

Nomen est omen. Wave is based on the form of a crashing wave that takes over the base and the panels of the bag. The zip fastener takes up this central theme as it runs from the front to the rear.

Thanks to the form of its base, the bag does not need any side sections. This gives it a slim, delicate appearance, but does not reduce its usable volume. In addition to the main compartment, there is a pouch on the inside for a mobile phone. The two larger bags also have an additional large compartment with a Velcro fastener.

The bag's two-tone finish accentuates the counter movements of the rear and front panels. The contrast between light and dark colours alludes to the play of light and shadow in a wave. Because the panels have been turned and the bag becomes narrower toward the top, Wave is very comfortable to carry.

The bag, which is produced in eco-friendly fashion in Berlin, has been developed to last at least ten years. Its outer skin is made of high-quality cowhide nappa leather from south Germany. This has been dyed all the way through, and so is colour-fast, non-fading and water-repellent. On the inside, Wave is lined with suede, which also ensures long life.

The bags, which are available in three sizes, are very versatile: the smallest bag is for going out on the town, the medium-sized one for everyday use, and the large one for documents, laptop, and small shopping errands.

Neolog OS
Armbanduhr

Wrist watch

Ambiente, Lifestyle
Silber

Ambience, lifestyle
Silver

Jury
Judges panel

Sehr interessanter, moderner Ansatz eines alten Themas: Die Darstellung der Zeit, die das Sanduhrenprinzip aufgreift, aber die Uhrzeit auf drei verschiedene Arten visualisiert. Uns gefällt auch sehr gut die innovative Verpackung, die die Formensprache der Uhr widerspiegelt. Sie ist stapelbar, was sich bei der Produktpräsentation im Verkauf positiv auswirkt. Mit dem Verkaufspreis von 120 Euro ist sie sehr attraktiv positioniert. Ein schöner Lifestyle-Artikel.

Very interesting modern approach to an old theme: a presentation of time that takes up the hourglass principle, but visualizes the time in three different ways. We also very much like the innovative packaging, which reflects the watch's formal language. It is stackable, which has a positive effect on product presentation at the point of sale. And its very attractively priced, at 120 euros. An attractive lifestyle article.

Hersteller
Manufacturer
Scopetime GmbH
D-22049 Hamburg

Design
Designer
Emamidesign
Arman Emami
D-10117 Berlin

Vertrieb
Distributor
Scopetime GmbH
D-22049 Hamburg

Neolog OS zeigt die Uhrzeit in drei verschiedenen Darstellungsformen. Per Knopfdruck kann nach Belieben gewählt werden: digitale Zahlenanzeige wie gewohnt oder aber Zeit als gruppierte Menge und nach dem Würfelaugenprinzip.

Ziel war, die Zeit, wie bei einer Sanduhr, als Menge zu zeigen. Dies veranschaulicht das Wachsen und Voranschreiten der Zeit viel deutlicher und intuitiver als die herkömmliche Darstellung mit Zahlen und Zeigern. Trotz der minutengenauen Darstellung wirkt der angezeigte Moment gegenständlich und ist auf das Wesentliche reduziert.

Beim Mengenprinzip wird die Zeit in Stunden, Zehner- und Einerminuten unterteilt. Die Einheiten werden als Balken dargestellt und zur schnelleren Wahrnehmung in Dreiergruppen gegliedert. Das Würfelaugenprinzip nutzt Bekanntes in neuem Kontext: Durch die vertraute Anordnung der Würfelaugen wird die Uhrzeit schnell erfasst.

Passend zu den unkonventionellen Darstellungsformen der Zeit besitzt das Uhrengehäuse eine sanduhrenähnliche, taillierte Form. Dadurch verschmelzen die zwei Druckknöpfe mit dem Gehäuse zu einer homogenen Einheit und lassen sich ohne hervorzustehen bedienen. Die wie aus einem Guss wirkende Formgebung besticht durch ihre organische Linienführung.

Die Uhr mit der futuristischen Optik besitzt ein kräftiges Silikonband, eine kontrastreiche organische Leuchtdiodenanzeige, Quarzsteuerung und ein Gehäuse mit doppelter Verglasung. Sie ist in verschiedenen Farben erhältlich.

Neolog OS gives the time in three different ways. At a touch of a button, the wearer can choose from a conventional digital display, grouped bars, or cubed pips.

The idea behind the watch was to present time as a quantity, as in an hourglass. This visualizes the progress and passage of time much more clearly and intuitively that if it is presented in the usual way with figures and hands. Despite being presented to the nearest minute, the moment that is displayed takes on concrete form, and is reduced to the essentials.

The grouped bars divide the time up into hours, and the minutes into tens and units. Each of these quantities is presented as a bar, grouped into blocks of three to make them easier to read. The cubed pips display takes a familiar idea and presents it in a new way. Once the different patterns of the pips have become familiar, the time can be told quickly.

Appropriately for this unconventional way of displaying the time, the housing of the watch has a tapered, hourglass-like shape. As a result, the two buttons merge with the housing to form a homogenous unit, and do not stick out. Seemingly cast from a single mould, the organic lines of the watch's shape are a striking feature.

Futuristic in appearance, the watch has a bold silicone strap, a high-contrast organic LED display, quartz control and a double-glass housing. It is available in various colours.

NEOLOG
NEOLOG
NEOLOG

Living Lounge
Schale

Fruit basket

Ambiente, Lifestyle
Silber

Ambience, lifestyle
Silver

Jury
Judges panel

Eine formal gelungene Neuinterpretation eines sehr bekannten Themas. Durch die silberne Oberfläche werden Gegenstände in der Schale sehr hervorgehoben. Wenn die Schale leer ist, ist sie ein schönes, ästhetisch gelungenes Objekt, einfach und gleichzeitig elegant. Die innere Schalenform ist elliptisch, fast organisch wie eine Eiform, was ein spielerisches Element neben der puristischen Form darstellt. Das macht die Schale interessanter und bildet einen schönen Kontrast.

A formally well executed re-interpretation of a very well known theme. As a result of the silver surface, the objects in the basket are given a lot of emphasis. And if the basket is empty, it is a beautiful, tasteful object – simple and elegant at the same time. The inner form of the basket is elliptical and almost organic like an egg, which is a playful element in addition to the purist form. This makes the basket more interesting, and creates an attractive contrast.

Hersteller
Manufacturer
WMF AG
D-73312 Geislingen

Design
Designer
Neunzig° Design
D-73240 Wendlingen

Vertrieb
Distributor
WMF AG
D-73312 Geislingen

Die außergewöhnliche Edelstahlschale bietet reizvolle Kontraste. Sie verbindet eine weiche organische Form im Inneren mit einer massiv anmutenden quadratischen Grundform außen. Rund trifft eckig – Gegensätze, die bei Living Lounge für Spannung sorgen.

Aus der archaischen Einfachheit des quadratischen Grundkörpers entsteht durch das Herauslösen eines skulpturhaften Freiformvolumens eine neue ästhetische und funktionale Gesamtheit. Die quadratische Außenkontur umschließt die weiche Vertiefung, die sich in den massiv wirkenden Block einschmiegt.

Die puristische und elegante Schale wirkt für sich als Designobjekt und setzt individuelle Wohnakzente. Wird sie befüllt, so scheinen alle in ihr platzierten Dinge gleichsam aus ihrem Kontext gelöst und freigestellt. Grund für dieses ungewöhnliche Phänomen ist die unterschiedliche Ausrichtung der Reflexionen auf den planen und den gewölbten hochglanzpolierten Oberflächen. Das führt zu einem Spannungsfeld von materiellem Volumen und immaterieller Wahrnehmung.

This unusual stainless steel basket is one of attractive contrasts. It combines a soft, organic form on the inside with a seemingly solid square basic form on the outside. Round meets square – in the case of Living Lounge, these opposites create a fascinating tension.

By allowing a sculptural free form to emerge from the archaic simplicity of the basic square, a new integrated whole is created, which is both tasteful and functional. The square outer contour wraps around the soft hollow, which nestles in the seemingly solid block.

Purist and elegant, the basket can stand on its own as a design object, and sets an individual accent in the home. If it is filled, then all the objects placed in it seem to be removed from their context and set free. The reason for this unusual phenomenon is the different direction of the reflections on the smooth and convex highly polished surfaces. This creates a tension between tangible volume and intangible perception.

Eva Solo Orchideentopf

Orchid pot

Ambiente, Lifestyle Silber

Ambience, lifestyle Silver

Jury
Judges panel

Absolut funktional. Die Orchidee nimmt sich nur das, was sie braucht. Man muss nur selten Wasser einfüllen. Hier haben wir zwei Gefäße in einem, die nicht mehr auseinanderfallen können. Nicht nur praktisch, sondern auch sehr ästhetisch. Gefällt uns ausgesprochen gut.

Absolutely functional. The orchid just takes what it needs. You probably only have to refill it once a month. Two pots in one that can no longer become separated. Not only practical, but also very tasteful. We like it very much.

Hersteller
Manufacturer
Eva Solo A/S
DK-2760 Måløv

Design
Designer
Tools Design
Henrik Holbæk,
Claus Jensen
DK-2400 Kopenhagen NV

Vertrieb
Distributor
Eva Solo A/S
DK-2760 Måløv

Die zarte und elegante Orchidee gehört mittlerweile zu den beliebtesten Zimmerpflanzen und passt gut zu einem stilvollen Ambiente. Und sie ist eine der pflegeleichtesten. Jedoch muss man beim Gießen aufpassen: Zuviel oder zuwenig Wasser schadet ihr. Ideal wäre, wenn sie sich selbst gießen und pflegen könnte.

Genau dafür hat nun Eva Solo einen selbstbewässernden Blumentopf entwickelt, der der Orchidee optimale Wachstumsbedingungen bietet. Über einen Nylonfaden am Boden des Topfes saugt die Pflanze nur so viel Wasser auf, wie sie braucht.

Der Eva Solo Orchideentopf besteht aus zwei Behältern, wobei in dem inneren Topf die Pflanze ihren Platz findet und der Äußere als Wasserreservoir dient. Durch eine seitliche Öffnung lässt sich das Reservoir einfach und bequem befüllen.

Der praktische Orchideentopf ist aus weißer, glasierter Fayence hergestellt und passt durch sein schlichtes, elegantes Design harmonisch in jedes Wohnambiente. Die runde, organische Form des Blumentopfes gibt der Orchidee den passenden Rahmen und bietet einen schönen Kontrast zu der eher skulpturalen Anmutung der Pflanze.

Delicate and elegant, orchids have become one of the most popular indoor plants, and are a fitting addition to any stylish interior. They are also one of the easiest plants to look after. However, care has to be taken when watering them. Too much or too little water will harm them. The best solution would be if they could water and look after themselves.

Eva Solo has had the same idea, and come up with a self-irrigating flower pot that offers orchids the best possible growing conditions. A nylon wick in the base of the pot allows the plant to draw up only as much water as it needs.

The Eva Solo orchid pot comprises two vessels, with the inner pot for the plant itself and the outer one acting as a water tank. An aperture at the side allows this tank to be filled simply and conveniently.

This practical orchid pot is made of white glazed faience. Its unostentatious, elegant design means that it harmonizes with any living style. The pot's round, organic form is the perfect setting for orchids, offering a beautiful contrast to the plant's rather sculptural air.

Gloria Design
Handsprüher

Hand-held sprayer

Ambiente, Lifestyle
Silber

Ambience, lifestyle
Silver

Jury
Judges panel

Ein schönes Produkt, weil es zwei Funktionen miteinander kombiniert: Sprühen und Gießen. Die wichtigsten Produktmerkmale sind der Quickverschluss, mit dem man das Produkt sagenhaft schnell auf und zu machen kann, und das Gewicht unten, wo das Wasser angesaugt wird, sodass man immer Wasser bekommt, egal in welcher Position man den Sprüher hält. Hervorragend auch die Doppelfunktion: Ausgießer und Trichter, der verhindert, dass Wasser beim Befüllen herunterläuft.

An attractive product, because it combines two functions. It can be used for spraying and watering. The best features of the product are the quick-release coupling, which allows it to be opened and closed incredibly quickly, and the weight in the base where the water is sucked up. This means water is always available, regardless of how the sprayer is being held. The dual function of spout and funnel is also excellent, keeping water from being spilled when re-filling.

Hersteller
Manufacturer
Gloria Haus- und Gartengeräte GmbH
D-58456 Witten

Design
Designer
Designaffairs GmbH
D-81671 München

Vertrieb
Distributor
Gloria Haus- und Gartengeräte GmbH
Vertriebs- und Marketingzentrale
D-89231 Neu-Ulm

Der Handsprüher bietet für den Einsatz in Wohnung und Wintergarten, auf Terrasse und Balkon einen Zweifachnutzen: Sprühen und Gießen. Die Pflanzen können sanft und schonend besprüht werden, ohne dass sie durch zu starken Wasserdruck geschädigt würden. Gerade auch in Innenräumen ist exakt dosiertes Sprühen dank der Präzisionspumpe sehr hilfreich. So kann ein feiner Wassernebel Morgentau simulieren, den die Pflanzen für gesundes Wachstum brauchen.

Praktisch ist die integrierte Ausgießöffnung, mit der Blumen und Pflanzen gegossen und gewässert werden können. Sie dient zugleich als großer Einfülltrichter, der einfaches Befüllen ermöglicht. Komfort bieten neben dem ergonomisch gestalteten Kopf und Taster auch die stufenlos verstellbare Düse für feinste Zerstäubung und der Quickverschluss, mit dem sich der Pumpkopf einfach und schnell abnehmen lässt. Durch die 360°-Sprühfunktion kann in ganz unterschiedlichen Positionen gesprüht werden.

Der Handsprüher, der in fünf verschiedenen Farben erhältlich ist, besticht nicht nur durch seinen hohen Ausstattungskomfort, sondern auch durch sein elegantes und zeitloses Design und seine weiche, runde Formgebung.

When used in the home or the conservatory, on the patio or balcony, this hand-held sprayer offers a dual benefit: it can be used for spraying and watering. Plants can be sprayed softly and gently, without damaging them as a result of too much water pressure. Especially in the home, it is very helpful that its precision pump allows the spray to be precisely metered. It can simulate a fine morning mist of morning dew, which plants need for healthy growth.

The integrated spout is very practical, allowing flowers and plants to be watered. The spout also serves as a large filling funnel, so that the sprayer can be easily refilled. The sprayer's many convenient features include an ergonomically shaped head and lever, an infinitely variable nozzle for the finest of sprays, and a quick-release coupling for removing the pump head simply and quickly. Its 360-degree spray function means it can be used in any position.

The hand-held sprayer, which is available in five different colours, is striking not only for its very convenient features, but also for its elegant and timeless design and soft, round shape.

SL22 Duell
Spardose

Money box

Ambiente, Lifestyle
Silber

Ambience, lifestyle
Silver

Jury
Judges panel

Eine entzückende Idee: Man hat zwei Einwurfschlitze für Geld, das heißt, man kann sich entscheiden, ob man lieber für das eine oder das andere spart. Oder es können zwei Personen damit sparen und sehen, wer mehr spart. Eine Idee, die es so am Markt noch nicht gibt. Sehr schön umgesetzt. Die Materialkombination von Keramik mit dem Silikonband finden wir gut gelungen.

A charming idea: you have two slits for money, so you can decide whether you want to save for one thing or the other. Or two people can save, and see who is saving more. An idea that has not been seen on the market before. Very well executed. We like the combination of porcelain and silicone band.

Hersteller
Manufacturer
Konstantin Slawinski
Housewarming objects
D-50827 Köln

Design
Designer
Ding 3000
D-30453 Hannover

Vertrieb
Distributor
Konstantin Slawinski
Housewarming objects
D-50827 Köln

SL22 Duell stellt die moderne Version des Sparschweins dar. Aus den zwei Sparschlitzen wird mit Hilfe der richtigen Beschriftung ein spannendes Duell. Einer gegen den anderen. Oder Kino versus Essengehen. Oder neue Tasche gegen Schuhe. Oder Konzert versus Theater.

Die Spardose für Sparduelle und Themensparen ist in der Mitte in zwei Hälften unterteilt. Jeder Cent zählt. SL22 Duell zeigt spielerisch an, wer vorne liegt oder welcher Wunsch zuerst erfüllt werden kann. Durch ihren runden Boden neigt sie sich auf die Seite, auf der mehr gespart wurde und heizt den Wettbewerb an. Damit macht Sparen Spaß.

Die Spardose besteht aus weißem Porzellan, das immer wieder beschriftet werden kann. Um die Dose, die oben offen ist, ist ein rotes Silikonband mit den beiden Sparschlitzen gespannt. Dieses wird, will man das Sparschwein »schlachten«, einfach heruntergezogen und der Wunscherfüllung steht nichts mehr im Wege.

SL22 Duell is a modern version of the piggy bank. Each of the two slits is given a name, and the result is an exciting duel. One against the other. Cinema against restaurant. New handbag against shoes. Concert against theatre.

This money box for saving duels and themed saving is divided into two halves. Every cent counts. SL22 playfully shows who is ahead, and which wish can be fulfilled first. With its rocking base, it leans toward the side into which more money has been saved, heating up the competition. It makes saving fun.

The money box is made of white porcelain that can be re-labelled again and again. A red silicone band into which the two saving slits has been cut runs around the money box, which is open at the top. When the time comes to break into the piggy bank, the band is simply pulled off, and nothing more stands in the way of your wishes.

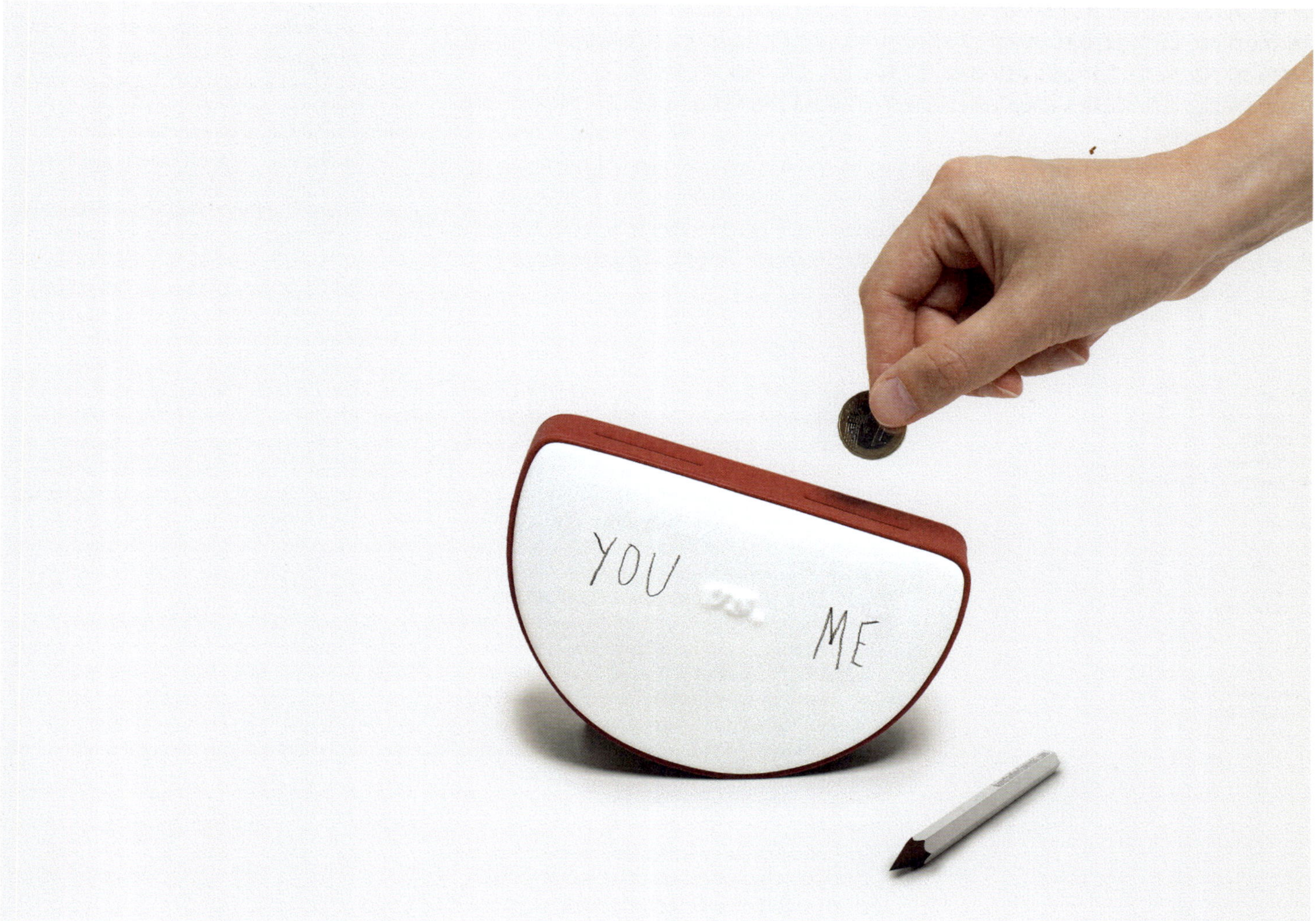
YOU
ME

Dyson Air Multiplier AMO2
Turmventilator

Tower fan

Jury
Judges panel

Ambiente, Lifestyle
Silber

Ambience, lifestyle
Silver

In seiner Grundfunktion überraschend gut: Ein starker Luftaustritt aus einem völlig unbewegten Element, ein klassisches Beispiel für die Kombination aus Innovation und Design. Und er ist leise. So muss Gestaltung sein: Das Weglassen von allem Überflüssigen. Er macht nur bewegte Luft.

Surprisingly good in its basic function. A powerful jet of air from an element that does not move at all – a classic example of the combination of innovation and design. And it is silent. Fundamentally, you can say this is how design should be: completely without frills. All it does is move the air.

Hersteller
Manufacturer
Dyson Ltd.
Malmesbury,
Wiltshire SN 16 ORP UK

Design
Designer
Werksdesign
In-house design
James Dyson

Vertrieb
Distributor
Dyson GmbH
D-50825 Köln

Nach der Produkteinführung des ersten Ventilators ohne Rotorflügel, des Air Multiplier AMO1, ist jetzt für große Räume ein neuer Air Multiplier Ventilator entwickelt worden: der Turmventilator AMO2. Mit einer Höhe von einem Meter und einer Breite von 19 Zentimetern besitzt er eine schlanke und geradlinige Form, die nur wenig Platz in Anspruch nimmt.

Die revolutionäre, patentierte Air Multiplier Technologie sorgt dafür, dass pro Sekunde 33 Liter Luft durch ein schraubenförmiges Antriebsrad angesaugt und um das 16fache verstärkt werden. Lediglich sieben Prozent der vom Ventilator erzeugten Luft fließen tatsächlich durch dieses Antriebsrad, 93 Prozent entstehen durch das »Einsaugen« der umgebenden Raumluft in den primären Luftstrom.

Die Luft wird dabei durch eine schmale Öffnung innerhalb des Luftrings beschleunigt. Die Schräge am Luftring ist wie eine Tragfläche geformt, um einen schnellen und starken Luftstrom aufzubauen. Dadurch wird ein Hochdruck-Luftstrom erzeugt, der über die tragflächenförmige Rampe fließt und so seine Richtung kanalisiert. Zusätzlich wird die umgebende Luft mit in den Luftstrom gezogen. Ähnliche Technologien werden auch bei Turboladern und Düsentriebwerken eingesetzt.

Der Air Multiplier AMO2 besitzt eine Fernbedienung, die magnetisch ist und gebogen geformt wurde, um sie direkt am Ventilator fixieren zu können. Mit dieser Fernbedienung oder direkt am Gerät kann AMO2 präzise und stufenlos eingestellt und nach oben und unten gekippt werden.

Following the launch of the first fan without blades, the Air Multiplier AM01, a new Air Multiplier fan has now been developed for large rooms: the AM02 tower fan. One metre high and 19 centimetres across, its slim, straight form means it takes up very little space.

Thanks to its revolutionary, patented Air Multiplier technology, 33 litres of air are sucked in per second by a screw-shaped drive wheel, and amplified 16 times over. Just 7 percent of the air generated by the fan actually flows through this drive wheel. The remaining 93 percent is created by sucking the surrounding air into the primary air stream.

A thin slit inside the annular aperture accelerates the air. The oblique surface of the annular aperture is formed like an airfoil in order to build up a fast and powerful jet of air. A high-pressure jet of air is created that, in flowing over this airfoil-shaped ramp, is channelled in a certain direction. At the same time, the surrounding air is drawn into the air flow. Similar technologies are used in turbochargers and jet engines.

The Air Multiplier AM02 comes with a remote control. Magnetic, and curved in shape, it can be fixed directly to the fan. Using the remote control or the controls directly on the fan, AM02 can be precisely and steplessly adjusted, and tilted upwards and downwards.

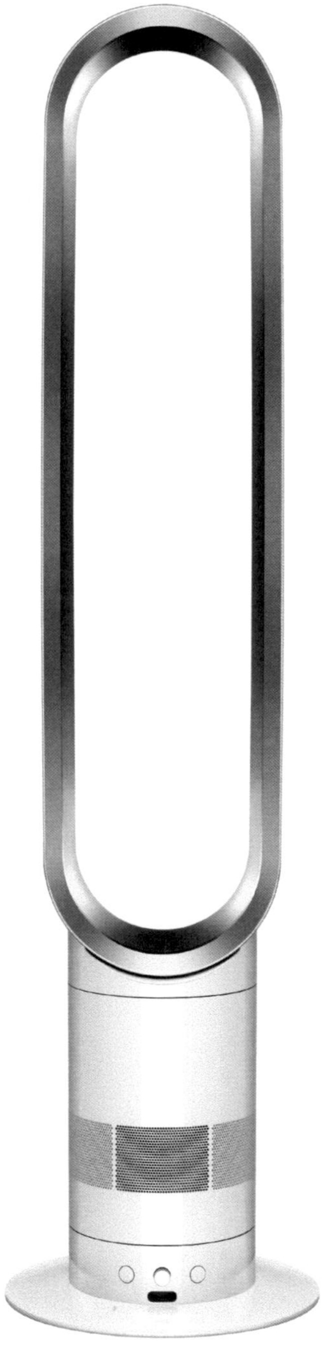

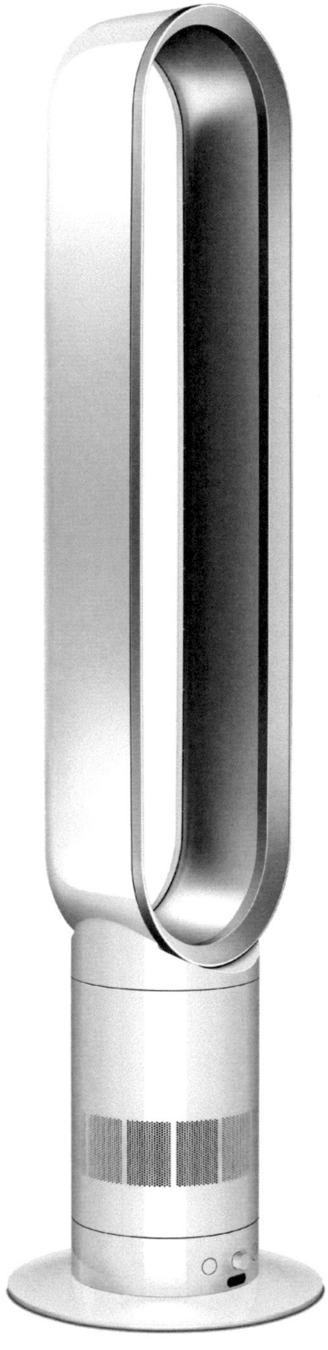

SL28 James
Garderobe James

James wardrobe

Jury
Judges panel

Ambiente, Lifestyle
Silber

Ambience, lifestyle
Silver

Formal gelungen, wie die Garderobe aus der Wand heraus wächst. Uns gefällt, dass man einen schönen Korb an der Wand assoziiert. Ein attraktives Designobjekt im Wohnambiente. Wir sehen sie nicht nur als Garderobe, sondern auch als Spontanablage. Man kann auch andere Dinge als Jacken und Mäntel aufhängen, vielleicht Schlüssel oder andere Utensilien.

What we found formally attractive about this was the way the wardrobe grows out of the wall. We like its association with a beautiful basket on the wall. An attractive designer object for the home. We don't see it just as a wardrobe, but also as somewhere to put things when you're in a hurry. Jackets and coats aren't the only things you can hang on it – perhaps keys as well.

Hersteller
Manufacturer
Konstantin Slawinski
Housewarming objects
D-50827 Köln

Design
Designer
Samuel Treindl
D-48155 Münster

Vertrieb
Distributor
Konstantin Slawinski
Housewarming objects
D-50827 Köln

Besitzt die Garderobe nun Haken oder Netz? James bietet beides und noch mehr. Das ungewöhnliche Produkt fungiert nicht nur als Garderobe, sondern zugleich auch als Ablage. Ein moderner Allrounder, der Platz sowohl für Jacke und Mantel bietet als auch Schal, Handschuhe und Mütze beherbergt.

Bei SL28 James wurden Kleiderhaken und das stabile Ablagenetz aus einem einfarbigen Stahlblech geschnitten. Netz und Haken scheinen aus der Wand herauszuwachsen. Damit wurde eine platzsparende Lösung für eine attraktive Garderobe geschaffen, die ideal in kleinen Wohnungen eingesetzt werden kann und zugleich mit ihrem modernen Design einen schönen Akzent setzt.

Die Garderobe wird in Deutschland aus pulverbeschichtetem Stahlblech hergestellt und ist in den Farben Schwarz oder Weiß zu erhalten.

Does this wardrobe have hooks or a net? James has both, and more. This unusual product acts not only as a wardrobe, but also as a rack. A modern all-rounder that offers room for jackets and coats, as well as for scarves, gloves and hats.

The hooks and net of SL28 James are cut out of a single-colour piece of sheet metal. The net and the hooks seem to grow out of the wall. The result is a space-saving solution for an attractive wardrobe. It is ideal for small apartments, and its modern design sets an attractive accent.

The wardrobe is made in Germany from powder-coated sheet steel, and is available in black or white.

C116 Modern Cubus
Klavier

Piano

Jury
Judges panel

Ambiente, Lifestyle
Silber

Ambience, lifestyle
Silver

Eine Neuinterpretation einer klassischen Struktur, die moderne Architektur und Linienführung, strengen Kubismus und Reduktion aufgreift und das Klavier in die Zukunft transferiert. Es begegnet der modernen Linienführung von Möbeln und nimmt sich zurück, es wirkt nicht wie ein Solitär. Viele Leute wollen nicht, dass das Klavier ein dominantes Möbelstück ist. Unter den Klavieren ist es ein Quantensprung.

A re-interpretation of a classic structure that takes up themes from modern architecture and lines, from strict cubism and reduction, and transposes the piano into the future. It responds to the modern lines of furniture. An unassuming piece of furniture, it does not give the impression of being a one-off piece. Many people do not want a piano to be a dominant piece of furniture. In piano terms, this is a quantum leap.

Hersteller
Manufacturer
Wilhelm Schimmel
Pianofortefabrik GmbH
D-38122 Braunschweig

Design
Designer
Werksdesign
In-house design
Sven Schneider

Vertrieb
Distributor
Wilhelm Schimmel
Pianofortefabrik GmbH
D-38122 Braunschweig

C116 Modern Cubus ist ein Klavier, bei dem alle Sinne angesprochen werden. Hier verbindet sich höchster Musikgenuss mit erstklassigem Design. Die klare und minimalistische Gestaltung in Verbindung mit neuen Fertigungstechniken und Materialien ermöglicht eine in der Welt der Klaviere außergewöhnliche Ästhetik.

Die klassische Konsolenbauweise betont die Vertikale und verleiht dem Klavier damit optisch mehr Größe. Die ungewöhnliche Materialkombination unterstreicht die wesentlichen Gestaltungselemente des Instruments. So bildet Aluminium einen spannungsvollen Kontrapunkt zum polierten Klavierlack. Das außergewöhnliche, puristische Design, die Reduktion auf das Notwendige und die strenge rechteckige Linienführung verleihen dem hochwertigen Klavier seinen ganz eigenen Charakter und lassen es zeitlos elegant wirken.

Das Klavier ist in Schwarz und in Weiß zu erhalten. Es besitzt ein durchgehendes Notenpult, der Deckel und die Klappe sind verschließbar. Praktische Griffe und stabile Rollen erleichtern den Transport.

C116 Modern Cubus is a piano that appeals to all the senses. It merges outstanding musical pleasure with first-class design. Its clear and minimalistic shape, combined with new manufacturing techniques and materials, results in an aesthetic that is rare in the piano world.

Its classic console construction emphasizes the vertical, giving the piano the impression of greater size. Its unusual combination of materials highlights the instrument's main structuring elements. For example, aluminium creates an exciting counterpoint to the polished piano lacquer. Its unusual, purist design, reduction to essentials and strict rectilinearity give this high-quality piano a character all of its own, as well as an air of timeless elegance.

The piano is available in back and white. It has a continuous music stand, and its lid and cover can be locked. Practical grips and stable casters make it easier to transport.

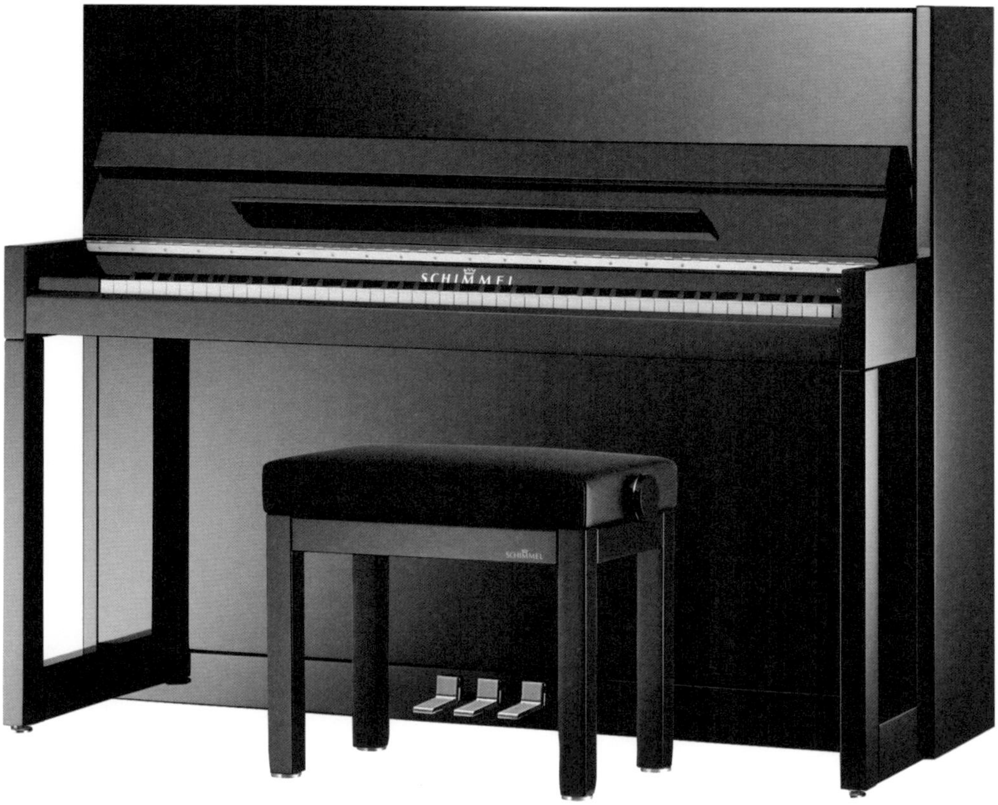
SCHIMMEL
SCHIMMEL

Hersteller
Manufacturer
Steng Licht AG
D-70327 Stuttgart

Design
Designer
Werksdesign
In-house design
Andreas Steng

Vertrieb
Distributor
Steng Licht AG
D-70327 Stuttgart

OPTIMAL LED STENG LICHT

Optimal-LED
LED-Downlight

LED downlight

Beleuchtung
Gold

Lighting
Gold

120 121

Optimal-LED
LED-Downlight

Beleuchtung
Gold

LED downlight

Lighting
Gold

Jury
Judges panel

Absolut reduziert, weniger ist kaum möglich. Es geht nur um das Licht. Alles ist hervorragend gestaltet, sogar der eigentlich nicht sichtbare Korpus. Selbst wenn die Leuchte nicht eingebaut würde, hätte sie immer noch eine extrem schöne Formensprache. Die Leuchte ist solide gemacht, ein hochwertiges Produkt. Einfach einzubauen.

Absolutely reduced, less is scarcely possible. Light is all that matters here. Everything has been beautifully designed, even the rear surface of the heat sink. Even if the luminaire were not surface mounted, it would still possess an extremely beautiful formal expression. The lamp is very well made, a high-quality product. Simple to install.

Hersteller
Manufacturer
Steng Licht AG
D-70327 Stuttgart

Design
Designer
Werksdesign
In-house design
Andreas Steng

Vertrieb
Distributor
Steng Licht AG
D-70327 Stuttgart

Die seit vielen Jahren am Markt erfolgreiche Optimal-Produktfamilie, bekannt durch den Einsatz im Objektbereich, ist um eine hochwertige und lichttechnisch hervorragende Einbauleuchte erweitert worden: Optimal-LED.

Sie ist mit Hochleistungs-LEDs neuester Generation ausgerüstet. Puristische Gestaltung und konsequente Umsetzung von technischem Know-how zeichnet den Reflektor-Einbaustrahler aus. Er bietet hinsichtlich der Farbtemperatur, der Farbwiedergabe, des hohen Lichtstroms und der langen Lebensdauer von 50.000 Stunden eine sinnvolle und energieeffiziente Alternative zu herkömmlichen Lichtquellen.

Beispielsweise ist der Lichtstrom der 18-Watt-Ausführung (1.000 Lumen) um circa zehn Prozent höher als eine 50-Watt-Niedervolt-Halogenlampe. Alle Optimal-LED-Downlights können wahlweise mit dieser Variante oder mit 9 Watt (580 Lumen) oder 13 Watt (800 Lumen) betrieben werden.

Außerdem wurden die Einbauleuchten mit hochwertiger Reflektortechnik im Abstrahlwinkel 20, 40 oder 60 Grad ausgestattet. Der Korpus wird aus einem massiven Aluminiumblock gefertigt und bietet dadurch ein hervorragendes Thermomanagement. Zu den LED-Downlights sind verschiedenste Zuschaltgeräte erhältlich – dimmbar oder nicht dimmbar.

Die Optimal-Systemfamilie, die mittlerweile aus Pendel-, Stab- und Flex- bis hin zu Objektleuchten besteht, wird neben dem Privatbereich besonders auch in Museen, Hotels, Restaurants, Galerien und im Ladenbau eingesetzt.

The Optimal product family, which is well known for its use in architectural lighting, has been successful in the market for many years. It has now been extended to include a high-quality surface luminaire that provides excellent light: Optimal-LED.

It is equipped with the latest-generation high-power LEDs. A purist design and the strict application of technical know-how are what make this reflector surface luminaire so special. With its colour temperature, colour reproduction, high luminous flux, and long service life of 50,000 hours, it offers a sensible and energy-efficient alternative to conventional light sources.

For example, the luminous flux of its 18-watt version (1,000 lumen) is roughly ten times higher than that of a 50-watt low-voltage halogen lamp. All Optimal-LED downlights can be operated either in this variant, or with a 9-watt (580 lumen) or 13-watt (800 lumen) light.

In addition, the surface luminaires feature high-quality reflector technology, with a beam angle of 20, 40 or 60 degrees. The body is made of a block of solid aluminium, which gives it excellent thermal management properties. Various switches are available in combination with the LED downlights, some of which offer a dimming function.

The Optimal family, which now comprises pendant lamps, tubular lamps, flex lights, and lamps for office buildings, is used in the private sphere, but especially in museums, hotels, restaurants, galleries and shops.

Modul Q V6
Deckeneinbauleuchte

Flush-mounted ceiling light

Jury
Judges panel

Beleuchtung
Silber

Lighting
Silver

Eine interessante, wirklich innovative Lösung. Das Material verbunden mit der satinierten Oberfläche ist sehr stimmig. Überzeugend die Flexibilität, dass man die Lichtelemente gegeneinander und in verschiedene Richtungen ausrichten kann. Die Leuchte hat etwas Schönes, Spielerisches und erinnert an Lego. Sie spricht emotional an. Das finden wir sehr gut.

An interesting, truly innovative solution. The way the material has been combined with the satin-finish surface makes a lot of sense. Impressively flexible, the way lights can be turned to face each other and set in different directions. The light has elegant, playful qualities, reminiscent of Lego. It has an emotional appeal. We like this a lot.

Modul Q V6 is a flat ceiling light that is equipped with revolutionary LED.next technology.

The light stands out for its reduced, functional design, its geometric shape and its clear, strict lines. Its sober, timeless design is further underscored by the light's transparent, satin-finish surface.

Unlike other flush-mounted ceiling lights, it is furnished with movable elements. The individual spot segments can be turned axially by 35 degrees in both directions.

The ceiling light is available with a beam angle of 5, 9 or 15 degrees. The 16-watt LED is equivalent to a 100-watt halogen lamp, and delivers more than 969 lumen.

Hersteller
Manufacturer
Nimbus Group GmbH
D-70469 Stuttgart

Design
Designer
Werksdesign
In-house design

Vertrieb
Distributor
Nimbus Group GmbH
D-70469 Stuttgart

Modul Q V6 ist eine flache Deckenleuchte, die mit der revolutionären LED.next-Technologie ausgestattet ist.

Die Leuchte zeichnet sich durch ihre reduzierte funktionale Gestaltung, ihre geometrische Formgebung und ihre klare, strenge Linienführung aus. Das sachliche und zeitlose Design wird durch die transparent satinierte Oberfläche der Leuchte noch unterstützt.

Im Vergleich zu anderen Einbaudeckenleuchten besitzt sie bewegliche Elemente. So können die einzelnen Spotsegmente axial um 35 Grad gedreht werden, und das in beide Richtungen.

Die Deckenleuchte ist wahlweise mit einem Abstrahlwinkel von fünf Grad, neun Grad oder fünfzehn Grad erhältlich. Die 16-Watt-LED entspricht einer 100-Watt-Halogenlampe. Und sie liefert über 969 Lumen.

Annex Ceiling
Deckenleuchte

Ceiling light

Jury
Judges panel

Beleuchtung
Silber

Lighting
Silver

Eine schöne, dekorative Leuchtenfamilie. Ausgesprochen ästhetisch: die Materialanmutung und das Spiel mit verschiedenen Transparenzen. Eine Leuchte, die auch im ausgeschalteten Zustand attraktiv ist.

An elegant, decorative family of lights. Exceptionally tasteful: the impression created by the materials and the play on different levels of transparency. A light that is also attractive when it is switched off.

Hersteller
Manufacturer
Serien Raumleuchten
D-63110 Rodgau

Design
Designer
Uwe Fischer
Industrial Design
D-60594 Frankfurt

Vertrieb
Distributor
Serien Raumleuchten
D-63110 Rodgau

Eine außergewöhnliche Gestaltung und Formensprache charakterisiert die Leuchte Annex: Sie ist pure Form. Ein reiner, geometrischer Körper steht in einer »Vitrine« aus mundgeblasenem Kristallglas. Die Form der Vitrine korrespondiert dabei mit dem zur Schau gestellten Exponat und bildet mit ihm eine formale Einheit. Gleichzeitig hat sie aber auch die Funktion des Schirmes inne und ist an der Decke befestigt – und zwar ohne störendes Detail.

Diese radikale Reduzierung ist möglich durch ein weißes, vulkanisiertes Silikonformteil, über das das Glas gestülpt wird.

Annex ist in drei Größen und mehreren Varianten erhältlich. Mit ihnen lassen sich differenzierte Lichtwirkungen für hohe und niedrige Räume erzielen. So strahlt die Leuchte mit dem konischen Reflektor aus Milchüberfangglas direkt nach unten und erhellt zugleich Raum und Decke, während sich dagegen mit einem Reflektor aus Aluminium Licht dramatischer inszenieren lässt. Eine dritte Variante mit einem Schirm aus Milchüberfangglas ohne Innenreflektor leuchtet den Raum gleichmäßig aus.

Annex gibt es nicht nur als Decken-, sondern auch als Pendelleuchte. Sie ist vielseitig einsetzbar von Bädern und Küchen über Hausflure bis hin zu Büros.

The Annex light is characterized by unusual design and formal expression. It is pure form. A pure, geometric body is held inside a »showcase« made of mouth-blown crystal glass. The form of the showcase corresponds with the exhibit on display, making up a formal unity. At the same time, however, it acts as a lampshade and is attached to the ceiling fixture without any distracting details.

This radical reduction is possible thanks to a white, vulcanized silicon moulding, over which the glass is slipped.

Annex is available in three sizes and several variants. They can be used to create different light effects in high- and low-ceilinged rooms. The light with its conical reflector of opal glass shines directly downward, illuminating the room and the ceiling at the same time, while the variant with an aluminium reflector allows more dramatic light arrangements to be created. A third variant with a shade of opal glass without an internal reflector provides even lighting for the entire room.

Annex is available not only as a ceiling light, but also as a pendant light. It can be used in many settings, from bathrooms and kitchens to hallways and offices.

Logotec LED
Strahler für Stromschiene

Spotlight for track lighting

Jury
Judges panel

Beleuchtung
Silber

Lighting
Silver

Dieser Strahler passt gut zur Philosophie von Erco, denn es geht um die Gestaltung des Lichts und nicht um die Gestaltung von Leuchten. Intelligent gemacht wegen des modularen Grundgedankens. Man hat verschiedene Projektionsoptiken, mit denen man unterschiedliche Lichtakzente und -stimmungen schaffen kann. Eine weitere Innovation ist der Einsatz von LED-Technologie, und das Wichtigste dabei: die ideale Verbindung von Lichtqualität mit Energieeinsparung.

It goes well with the Erco philosophy, since this spotlight is about designing light and not designing lamps. Cleverly done with its basic modular idea. Various projection optics are available, allowing different highlights and moods to be created. The use of LED technology is a further innovation. Most importantly, it offers the ideal combination of light quality and energy saving.

Hersteller
Manufacturer
Erco GmbH
D-58507 Lüdenscheid

Design
Designer
Werksdesign
In-house design
Erco Design Team

Vertrieb
Distributor
Erco GmbH
D-58507 Lüdenscheid

Logotec LED ist das erste Strahlerprogramm von Erco, das ausschließlich für den Einsatz von LEDs optimiert wurde. Schon jetzt ist es mit seiner modernen Linsentechnik ein gleichwertiger Ersatz für alle Strahler mit herkömmlichen Reflektoren und Niedervoltlampen bis 100 Watt.

Die flache Form der Leuchte ist charakteristisch für den Einsatz der neuen LED-Technologie und trägt zum optimalen Wärmemanagement der Hochleistungs-LEDs bei. Die Leuchte zeichnet sich durch ihre kompakte Bauform aus, in der bereits das Betriebsgerät für die LEDs integriert ist.

Um gerichtetes Licht zu erzeugen, werden eigens entwickelte Projektionsoptiken eingesetzt. Sechs verschiedene Abstrahlcharakteristiken stehen bei dem Strahler zur Verfügung und bieten damit eine außergewöhnliche Vielfalt an Lichtcharakteristiken für die professionelle Lichtplanung. Die wechselbaren Spherolitlinsen bestimmen die Lichtverteilung: von enger Beleuchtung, die Akzente setzt, über breite Lichtkegel, um Flächen und Raumzonen zu beleuchten zum Beispiel bei der Warenpräsentation, bis hin zur vertikalen Beleuchtung. Damit lassen sich komplexe Beleuchtungsaufgaben durchgängig lösen.

Über den Bügel können alle Strahler des Programms geschwenkt und gedreht werden. Logotec ist auch als Einbaustrahler erhältlich.

Logotec LED is the first Erco range to have been optimized solely for the use of LEDs. Even now, its modern lens technology makes it a suitable alternative to any spotlights with conventional reflectors and to low-voltage lamps of up to 100 watts.

The lamp has the flat shape that is typical of new LED technology – a shape that also contributes to the optimum heat management of the high-performance LEDs. The lamp is strikingly compact, despite the fact that the control unit for the LED is integrated into it.

To generate directional light, the spotlight uses projection optics developed in the company. There is a choice of six different radiation characteristics, which means that there is an extraordinary variety of lighting moods for professional lighting design. The interchangeable Spherolit lenses determine how light is dispersed: from a narrow beam for highlights to a broad beam to illuminate spaces and room zones – for example when presenting goods – to vertical lighting. Complex lighting tasks can be solved using the same spotlights.

The bracket allows all the spotlights in the range to be swivelled and turned. Logotec is also available as a flush-mounted spotlight.

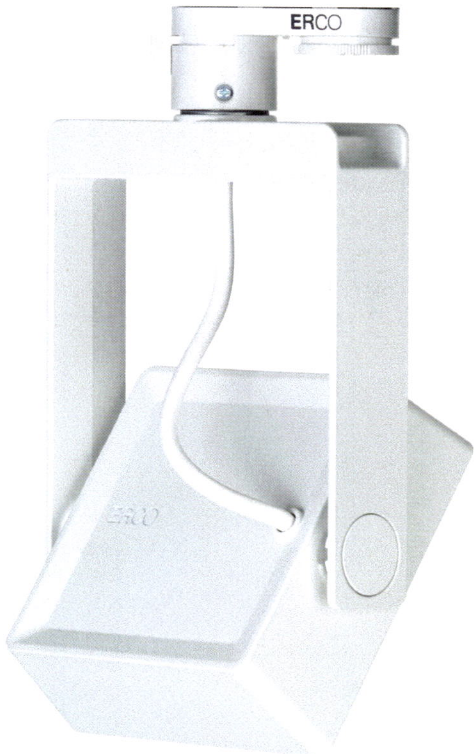

Pico
Handleuchte

Portable lamp

Beleuchtung
Silber

Lighting
Silver

Jury
Judges panel

Günstiges Produkt für viele individuelle Situationen. Multifunktionales Gerät für den Einsatz überall dort, wo es keinen Strom gibt: in Dritte-Welt-Ländern, im Campingbereich, in der Werkstatt oder bei Autopannen. Man kann sie aufhängen, sie hat 55 Stunden Leuchtdauer und lässt sich über ein Solarmodul aufladen. Über USB lassen sich zusätzlich andere Geräte, wie zum Beispiel das Handy, laden.

Reasonably priced product for many different situations. Multifunctional device for use wherever there is no electricity: in Third World countries, when camping, in repair shops, or when the car breaks down. It can be hung anywhere, lasts for 55 hours, and can be recharged using a solar module. A USB port allows other devices, such as a mobile, to be charged.

Hersteller
Manufacturer
Phocos AG
D-89077 Ulm

Design
Designer
Becker GmbH
D-89520 Heidenheim

Vertrieb
Distributor
Phocos AG
D-89077 Ulm

Pico ist dort einsetzbar, wo kein Stromnetz zur Verfügung steht, wie zum Beispiel auf dem Lande in der Dritten Welt, im Gebirge, in der Wüste, im Garten oder beim Camping. Dafür verfügt die multifunktionale LED-Solarleuchte über ein kleines, aber komplettes Stromversorgungssystem.

Angeschlossen an ein kleines Solarmodul, steuert der integrierte Laderegler den Stromfluss zum integrierten Akku. Zudem kann die Leuchte auch an einer Autobatterie oder an einer Steckdose geladen werden. Andererseits lassen sich Mobiltelefone oder MP3-Player mit dem passenden Strom versorgen.

Einen Tag geladen, liefert Pico über zwölf Stunden Licht mit einer Leistung von 50 Lumen. Bei Helligkeitsstufe eins kann sie sogar 55 Stunden betrieben werden. Mit ihrem hochwertigen, widerstandsfähigen und funktionalen Design ist Pico vielseitig einsetzbar: als Decken-, Wand- oder Handleuchte. Dank des gekapselten Gehäuses (IP 65) ist auch der Betrieb in staubigen oder feuchten Umgebungen wie beispielsweise am Strand möglich. Nachts ist Pico leicht zu finden, denn der phosphoreszierende Gummischutzring leuchtet im Dunkeln.

Vorbild für die Form der Leuchte war der Flaschenkürbis, der in Afrika beispielsweise auch als Gefäß oder Resonanzkörper im Musikinstrumentenbau eingesetzt wird. Die organische Form zeichnet sich durch ihre mechanische Stabilität aus und hat einen kleinen Nebeneffekt: Die Leuchte lässt sich als kleine Trommel zweckentfremden – mit erstaunlich gutem Klang.

Pico can be used wherever there is no power grid – in rural areas in the Third World, for example, in the mountains, the desert or in the garden, or when camping. This multifunctional solar-powered LED lamp has a small but complete electricity supply system.

When it is connected to a small solar module, the charging unit integrated in the Pico controls the flow of electricity to the integrated rechargeable battery. In addition, the lamp can also be charged from a car battery or from a mains electricity socket. At the same time, it can supply mobile phones and MP3 players with the electricity they need.

After charging for a day, Pico will deliver 50 lumen of light for more than 12 hours. At its lowest brightness setting, it will run for as long as 55 hours. Featuring a design that is high-quality, tough and functional, Pico is very versatile. It can be used as a ceiling, wall-mounted or portable lamp. Thanks to its encapsulated housing (IP 65), it can also be operated in dusty or damp environments – on a beach, for example. Pico is easy to find at night, since its phosphorescent protective rubber ring shines in the dark.

Its shape is based on a gourd, which in Africa is also used as a vessel, for example, or as a soundbox for musical instruments. This organic shape is mechanically robust, and there is an added effect: the lamp can also be used as a small drum – with an astonishingly good sound.

LED LENSER
M8
000001

LED LENSER® M8
LED Taschenlampe

LED pocket torch

Jury
Judges panel

Beleuchtung
Silber

Lighting
Silver

Diese Taschenlampe besitzt optimale Charakteristiken: Sie leuchtet sehr weit und ist besonders intensiv, sie ist handlich, klein, kompakt und leicht, sodass man sie in der Tasche verstauen kann. Eine zusätzliche interessante Funktion: sie ist wie ein Kamera-Teleskop-Auszug fokussierbar. Und sie ist äußerst einfach zu bedienen.

This pocket torch has excellent qualities: its beam distance is very long and especially intense, it is handy, small, compact and light, so that it can be kept in a pocket. An additional interesting function is the way it can be focused like the zoom lens of a camera. And it is extremely easy to use.

Hersteller
Manufacturer
Zweibrüder
Optoelectronics GmbH
D-42699 Solingen

Design
Designer
Werksdesign
In-house design
Rainer Opolka,
Andre Kunzendorf

Vertrieb
Distributor
Zweibrüder
Optoelectronics GmbH
D-42699 Solingen

Dank ihrer zwei temperaturbeständigen CR-Lithium-Batterien als Kraftquelle ist die M8 der ideale Begleiter beim Gipfel-Besteigen, bei der Erkundung von Wüsten oder der Wanderung durch Schnee und Eis.

Die Taschenlampe zeichnet sich durch ihr zeitloses Design und ihre funktionale Ergonomie aus. Die integrierte Smart-Light-Technologie ermöglicht es, wahlweise entweder Konstantstromregelung oder energieoptimierten Energy Saving Mode zu nutzen.

Je nach persönlichem Einsatzzweck bieten drei verschiedene Lichtprogramme mit bis zu neun verschiedenen Lichtfunktionen zahlreiche Möglichkeiten, sie einzusetzen. Diese reichen vom Easy-Programm für die gelegentliche Nutzung über das Professional-Programm mit vielfältigen Funktionen bis hin zur effektiven Selbstverteidigung mit Hilfe des Defense-Programms.

Die Reichweite der M8 ist eine der besten ihrer Klasse.

Herausragende Möglichkeiten innerhalb der fokussierbaren LED-Technik bietet das patentierte Fokussiersystem – Speed Fokus verbunden mit der Möglichkeit zur Schnellfixierung – Fast Lock. Besonders komfortabel ist, dass sich beide einhändig bedienen lassen.

Thanks to its power source, two temperature-resistant CR lithium batteries, the M8 is the ideal companion for mountaineering, exploring the desert and expeditions in the snow and ice.

The pocket torch is striking for its timeless design and functional ergonomics. Its integrated smart-light technology allows the user to choose between constant-current or energy-saving mode.

Depending on individual requirements, three different light programmes with up to nine different light functions offer many different possible uses. They range from the »Easy« programme for occasional use, to the »Professional« programme with a variety of functions, to the »Defence« programme with a strobe for effective self-defence.

The range of the M8 is among the best in its class.

When it comes to focusing the LED light, the patented »Rapid Focus« function is an excellent feature, combined with the »Fast Lock« function for rapidly setting a selected focus. What is especially convenient about these features is that they can be operated single-handedly.

LED LENSER® X14
LED Taschenlampe

LED pocket torch

Jury
Judges panel

Beleuchtung
Silber

Lighting
Silver

Ein funktionales Produkt, das einfach zu bedienen ist. Es hat zwei Lichtquellen, die binokular funktionieren und damit ein großes Streulicht erzeugen. Die Taschenlampe schafft es, durch eine einfache Bedienung diese beiden Lichtquellen auf einen Punkt zu einem intensiven Spotlight zu fokussieren. In der Form haben wir solch eine Leuchte noch nicht gesehen.

A functional product that is easy to use. It has two light sources that work binocularly, and this create a large scatter of light. Because the torch is so easy to use, these two light sources can be focused on one point, creating an intense spotlight. We have never seen a torch in a form like this before.

Hersteller
Manufacturer
Zweibrüder
Optoelectronics GmbH
D-42699 Solingen

Design
Designer
Werksdesign
In-house design
Rainer Opolka,
Yunzhou Wu

Vertrieb
Distributor
Zweibrüder
Optoelectronics GmbH
D-42699 Solingen

Ein ungewöhnlicher Lampenkopf mit zwei Lichtquellen ist das hervorstechendste Merkmal der X14. Sie gehört zu der Xtreme-Serie von LED LENSER® und besticht durch ihre beeindruckende Lichtleistung.

So strahlen bis zu 480 Lumen aus dem Kopf der Lampe und vereinigen sich zu einem fokussierbaren, kreisrunden Lichtkegel. Drei Lichtprogramme mit bis zu acht Lichtfunktionen, Smart Light Technology, ermöglichen die individuelle Konfiguration der Lampe.

Durch diese verschiedenen Programme ist die Lampe vielseitig einsetzbar. So ist sie für Sport und Hobby ebenso geeignet wie für Sicherheitsdienste oder für den taktischen Einsatz.

What is immediately striking about the X14 is its unusual torch head with two light sources. The torch belongs to the LED LENSER® Xtreme series, and stands out for its impressively powerful light beam.

The torch head produces up to 480 lumen of light, which can be focused into a circular cone of light. Thanks to smart-light technology, with three lighting programmes and up to eight lighting functions, the torch can be configured individually.

Thanks to these diverse programmes, the lamp can be used for a variety of purposes. It is as suitable for sports and leisure as it is for security services and tactical missions.

LED LENSER X14

Minox BN 7X50 DC
Fernglas

Binoculars

Kommunikation,
Audio / Video, Optik
Silber

Communication,
audio / video, optical equipment
Silver

Jury
Judges panel

Ein sehr schönes, funktionales Fernglas, das einfach einzustellen ist. Für große und kleine Augenabstände sowie für große und kleine Hände geeignet. Es enthält einen Kompass und mehrere andere Funktionen, die man beim Segeln miteinander kombinieren kann. Es ist wasserfest. Das weiße Fernglas spiegelt das weiße Segelimage wider. Ein gelungenes Produkt.

Very elegant, functional binoculars that are simple to adjust. Suitable for large and small distances between the eyes, as well as for large and small hands. It contains a compass and several other functions that can be combined when sailing. They are waterproof. The white binoculars are reminiscent of the yachting world. A well executed product.

Hersteller
Manufacturer
Minox GmbH
D-35578 Wetzlar

Design
Designer
Volkswagen Design
D-38436 Wolfsburg

Vertrieb
Distributor
Minox GmbH
D-35578 Wetzlar

Mit dem innovativen BN 7X50 DC beginnt eine neue Generation von Wassersport-Ferngläsern. Es besitzt richtungsweisende Vorteile gegenüber den bis heute verfügbaren Navigations-Ferngläsern mit herkömmlicher analoger Kompass-Technologie.

Revolutionär und einzigartig ist beim BN 7X50 DC der digitale Kompass im Zentrum des Sehfeldes, der vollständig im Gehäuse integriert ist. Das beobachtete Objekt sowie die Kompassinformationen bleiben dabei immer im Mittelpunkt der Aufmerksamkeit. Die Kompassanzeige kann entweder über Center Mark oder als Ansicht mit digitaler Strichplatte gewählt werden. Ansonsten bleibt das brillante Sehfeld jederzeit unbeeinträchtigt und frei von allen Markierungen und Anzeigen.

Das Fernglas lässt sich intuitiv bedienen: Per Knopfdruck erhält der Anwender im Bruchteil einer Sekunde alle wichtigen Skipper-Informationen digital, neben der Kompassanzeige mit automatischer Tilt-Kompensation – Anzeige des Neigungswinkels –, ein Barometer mit Luftdruckanzeige, eine Tilt-Funktion zur Höhenbestimmung von Objekten oder eine Stoppuhr direkt im Zentrum des Sehfelds eingeblendet und verliert somit das Wesentliche nie aus den Augen.

Weitere Vorteile des wasserdichten Fernglases, das es in Schwarz und Weiß gibt, sind das robuste Kunststoffgehäuse, die griffige Gummiarmierung und drehbare Augenmuscheln. Mit seiner vorbildlichen Ergonomie, hohen Funktionalität und herausragenden, hoch präzisen Optik erfüllt es die extremen Anforderungen auf hoher See und macht es zu einem komfortablen Begleiter an Bord.

The innovative BN 7X50 DC marks the start of a new generation of binoculars for water sports. They have groundbreaking advantages over the navigation binoculars that have been available up to now, with their conventional analogue compass technology.

What is revolutionary and distinctive about the BN 7X50 DC is their digital compass in the centre of the field of vision. The compass is completely integrated in the housing. The object under observation, as well as the information from the compass, always stay firmly at the centre of attention. The compass display can be shown either at centre mark or as a digital reticle. For the rest, the brilliant field of view remains unobstructed at all times, free of any marks or displays.

The binoculars can be operated intuitively: at the push of a button, the user receives all the important information a skipper needs in a fraction of a second. Apart from the compass display, the following information is projected directly into the centre of the field of vision: automatic tilt compensation (showing the angle of tilt), atmospheric pressure (via a barometer), a tilt function for determining the height of objects, and a stop watch. All this without losing sight of what is important.

Other advantages of the waterproof binoculars, which are available in black and white, are the sturdy plastic housing, the non-slip rubber armouring and the twisting eyecups. Its exemplary ergonomic features, high level of functionality and outstanding, highly precise optics mean that they fulfil the exacting requirements found on the high seas, making them a convenient shipmate.

MINOX

USB-Clip
Speichermedium

Storage medium

Kommunikation, Audio / Video, Optik Silber

Communication, audio / video, optical equipment Silver

Jury
Judges panel

Ein bekanntes Produkt neu interpretiert und neu kombiniert: Speichermedium mit der Zusatzfunktion Clip oder Büroklammer. Einfach, klein. Das Schöne ist, jeder kennt eine Büroklammer. Und mit diesem Gedanken wird hier gespielt. Ein Produkt, das schmunzeln lässt. Und auf diese Weise bleibt es im Gedächtnis. Ein Zusatznutzen: Es kann gebrandet werden und es ist nicht teuer. Ein schönes Give-away.

A re-interpretation and re-combination of a familiar product. A storage medium with an additional paperclip function. Simple and small. What's good about it is that everyone knows what a paper clip is. And this product plays with this idea. A product that makes you smile. And then you remember it. And there is a further plus point: it can be branded, it isn't expensive. A nice give-away.

Hersteller
Manufacturer
Freecom Technologies GmbH
D-10829 Berlin

Design
Designer
Emamidesign
Arman Emami
D-10117 Berlin

Vertrieb
Distributor
Freecom Technologies GmbH
D-10829 Berlin

Täglich werden millionenfach elektronische Speichermedien wie CDs, DVDs oder USB-Sticks mit Daten gefüllt und mit der Post verschickt. Dank der fortschreitenden technologischen Entwicklung werden USB-Sticks im Gegensatz zu CDs oder DVDs immer kleiner und sind damit einfacher und kostengünstiger zu versenden. Allerdings besteht die Gefahr, dass sie aufgrund ihrer Größe leicht verloren gehen.

Ganz anders bei dem USB-Clip. Er lässt sich einfach an jede Art von Unterlage – wie Visitenkarten, Briefpapiere oder Broschüren – klemmen. Drei kräftige Lamellen sorgen für die notwendige Haftung. Hinzu kommt, dass er die Unterlagen perfekt zusammenhält und somit analoges sowie digitales Material im wahrsten Sinne des Wortes optimal miteinander verbindet. Um sie vor Kratzern zu schützen, sind die empfindlichen Kontakte auf der Innenseite positioniert.

Der Adressat zieht den USB-Clip einfach von Visitenkarte oder Brief ab und steckt ihn in einen beliebigen USB-Anschluss seines Computers. Eine auffällige Beschriftung zeigt ihm an, in welcher Position der USB-Clip mit dem Computer verbunden werden muss und welche Speicherkapazität das Medium besitzt. Die gesamte äußere Seite kann mühelos bedruckt werden und eignet sich somit hervorragend für Logos und Werbebotschaften.

Es ist eine Frage der Zeit, bis portable Speichermedien ähnlich wenig kosten wie Büroklammern. In Zukunft könnte dann der Einsatz von Büroklammern durch kostengünstige und wiederverwertbare USB-Clips sinnvoll ergänzt werden.

Every day, millions of electronic storage media – CDs, DVDs and USB sticks – are filled with data and sent by mail. Thanks to technological progress, USB sticks are becoming smaller and smaller – unlike CDs and DVDs. This makes it simpler and cheaper to mail them. However, there is a danger that they might get lost because of their small size.

But not the USB-Clip. It can simply be clipped to any kind of document – business cards, letters or brochures. Three strong slats make sure it stays firmly put. In addition, it keeps documents together perfectly, and thus literally brings together analogue and digital content in the best possible way. To protect them against scratching, the sensitive contacts are positioned on the inside.

The recipient simply removes the USD-Clip from the business card or letter and inserts it into any of the USB ports on his computer. Clear labelling shows which way round the USB-Clip should be held when inserting it into the computer, as well as what capacity the stick has. The entire outer surface can be printed without difficulty, and so is eminently suitable for logos and advertising slogans.

It is only a matter of time before portable storage media cost as little as paper clips. In the future, it would then make sense to use inexpensive, reusable USB-Clips as well as paper clips.

TOP
GB 2
GB 2
2
2

just|mobile
just|mobile
just|mobile

AluPen™
Bedienstift

Stylus

Kommunikation,
Audio / Video, Optik
Silber

Communication,
audio / video, optical equipment
Silver

Jury
Judges panel

Funktional, schön, minimalistisch, liegt gut in der Hand. Es gibt genügend Anwendungen auf einem iPad, die mit einem Stift besser zu bedienen sind. Jeder erkennt: »das ist ein Stift, mit dem kann ich auf dem iPad zeichnen«. Wunderbar ist, dass man auf alte Gewohnheiten zurückgreift, das heißt einen Griffel in der Hand halten, das kann jeder, ob jung oder alt. Es ist das Bild eines Bleistiftes. Das finden wir sehr gelungen.

Functional, elegant, minimalistic, sits well in the hand. There are plenty of iPad applications that are easier to use with a stylus. Everyone knows »this is a pencil I can use for drawing on my iPad«. It's wonderful the way you can revert to old habits – holding a stylus in your hand is something all of us can do, young and old. It's the image of a pencil. We like this a lot.

Hersteller
Manufacturer
Just Mobile
D-12159 Berlin

Design
Designer
Tools Design
Henrik Holbæk,
Claus Jensen
DK-2400 Kopenhagen NV

Vertrieb
Distributor
Just Mobile
D-12159 Berlin

AluPen™ ist ein massiver Stylus, der die vertraute Form eines klassischen Bleistifts auf moderne Weise neu interpretiert. So besitzt der Bedienstift ebenfalls eine sechseckige geometrische Grundform, ist jedoch aus hochwertigem Aluminium hergestellt. Mit ihm lassen sich iPad, iPhone und iPod touch optimal und einfach bedienen.

Die ultrasofte patentierte Spitze aus Gummi garantiert feinste Kontrolle und lässt Schreiben und Zeichnen zu einem Erlebnis werden. Dementsprechend ist AluPen™ der optimale Partner für Mal- und Zeichen-Apps.

Der Bedienstift zeichnet sich aber nicht nur durch seine hohe Funktionalität der einfachen digitalen Eingabe, sondern auch durch sein ikonisches Design aus. Durch das Vorbild Bleistift bekommt AluPen™ einen großen Wiedererkennungswert und erfüllt die Erwartungshaltung so, wie jeder gelernt hat zu malen und zu schreiben.

Die Farbpalette, in der der Bedienstift erhältlich ist, lässt zudem an Buntstifte denken: neben Silber und Schwarz auch Gelb, Rot, Blau und Grün.

AluPen™ is a robust stylus that gives a modern re-interpretation of the familiar shape of the classic pencil. The stylus has the same basic hexagonal section, but is made of high-quality aluminium. It is the perfect tool for working with an iPad, iPhone and iPod touch.

The ultra-soft patented rubber tip guarantees precise control, making writing and drawing into a whole new experience. This makes AluPen™ the ideal partner for painting and drawing apps.

The stylus's highly functional property is not its only striking feature, however. Its design is also iconic. Being modelled on a pencil, AluPen™ has a high level of recognizability and satisfies users' expectations by allowing them to paint and write the way they learned.

The range of colours available for the stylus is also reminiscent of crayons: apart from silver and black, it is available in yellow, red, blue and green.

Slide™
Ständer für iPad

iPad stand

Kommunikation, Audio / Video, Optik Silber

Communication, audio / video, optical equipment Silver

Jury
Judges panel

Das Slide™ ist eine neue Interpretation für einen iPad-Ständer, der nicht viel vom Produkt wegnimmt, das Produktdesign nicht unterbricht oder stört. Die große, schwere Gummirolle ermöglicht es, das iPad in jedem Winkel zu positionieren. Gut gelöst finden wir auch, dass sich die Gummirolle integrieren und damit leicht transportieren lässt, indem sie in das zylinderförmige Ende hineingeschoben wird.

Slide™ is a new interpretation of an iPad stand. It does not detract from the product at the front, and so does not interrupt or disturb the iPad's product design. The large, heavy rubber cylinder allows the iPad to be positioned at any angle. We also like the solution for transporting the rubber cylinder: simply slide it into the tubular support.

Hersteller
Manufacturer
Just Mobile
D-12159 Berlin

Design
Designer
Tools Design
Henrik Holbæk,
Claus Jensen
DK-2400 Kopenhagen NV

Vertrieb
Distributor
Just Mobile
D-12159 Berlin

Slide™ ist ein einfacher und innovativer Ständer für das iPad, der universell einsetzbar ist.

Hergestellt wird er aus einem einzigen Stück hochwertigen Aluminiums. Der Clou ist die rutschfeste Gummirolle. Diese wird auf der Oberfläche des Slide™ platziert und das iPad dagegen gelehnt. Um den passenden Blickwinkel einzustellen, wird die Rolle einfach verschoben. Die Gummirolle positioniert das iPad sowohl im Hoch- als auch im Querformat optimal und bietet sichere Unterstützung beim Tippen, Wischen oder Spielen – und das in jeder Lage und jedem Winkel.

Wird der Ständer gerade nicht gebraucht oder an einen anderen Ort transportiert, wird die Gummirolle ganz einfach im zylindrisch geformten Ende des Slide™ untergebracht.

Seine klare funktionale Formgebung korrespondiert mit dem ästhetischen Design des iPad. Die puristische Gestaltung ist auf das Wesentliche reduziert und besticht durch ihre Eleganz.

Slide™ is a simple and innovative stand for an iPad. It can be used universally.

It is made out of a single piece of high-quality aluminium. Its ingenious feature is its non-slip rubber cylinder. This is placed on the surface of Slide™, and the iPad is leaned against it. To set the desired angle, the cylinder is simply slid into the right position. Whether in portrait or landscape format, the rubber cylinder places the iPad in the optimum position, offering secure support when typing, swiping or gaming – in any position and at any angle.

If the stand is not in use or is being transported, the rubber cylinder is simply stowed away in the tubular support of Slide™.

Its clear, functional form harmonizes with the iPad's tasteful design. Its purist shape is reduced to the bare essentials, and is outstandingly elegant.

iPad

Jury
Judges panel

Moonlet
iPad Accessoire

**Kommunikation,
Audio / Video, Optik
Silber**

iPad accessory

**Communication,
audio / video, optical equipment
Silver**

Was uns überzeugt, ist die minimalistische Verpackung des iPad und der erlebte Sturz aus einem Meter Höhe. Und es ist nichts passiert. Bei Menschen, die das iPad zum Spielen benutzen, liegt es besser in der Hand. Wir finden Moonlet sehr gelungen, weil es einfach nur aus vier Kugeln besteht. Fertig. Viel minimalistischer geht es nicht.

What especially impresses us is the minimalistic packaging of the iPad, and seeing the iPad drop from a height of one meter. The iPad wasn't damaged at all. And if people use the iPad as a toy, it fits better in the hand with these Moonlets that without them. This is what makes the product so good – the fact that it's just four balls. And that's all. You can't get much more minimalistic than that.

Hersteller
Manufacturer
Aturon GmbH
CH-9323 Steinach

Design
Designer
Werksdesign
In-house design
Marcus Höglmeier

Vertrieb
Distributor
Aturon GmbH (Europa)
CH-9323 Steinach und
Hudson Vandam LLC
(USA / Kanada)
New York, NY 10013

So simpel wie clever: Moonlet schützt das iPad und ergänzt dessen Gestaltung auf schöne Art und Weise. Aus der Schweiz kommt ihr Design und in Deutschland werden sie mithilfe modernster, patentierter Technologie gefertigt.

Die »kleinen Monde« werden einfach auf die Ecken des Tablet Computers gesteckt. So heben sie ihn von der Arbeitsfläche ab, wodurch Kratzer und unerwünschte Beschädigungen beispielsweise durch verschütteten Kaffee vermieden werden.

Außerdem ermöglichen sie in jedem Winkel eine stabile Handhabung beim Tippen und Arbeiten am iPad dank des rutschfesten Materials – sie bestehen aus einer hochwertigen Elastomer- / Gummimischung. Die kleinen Gummibälle, die in Schwarz und Weiß erhältlich sind, liegen sehr gut in der Hand, die angenehm samtige Oberfläche bietet dabei optimale Griffigkeit.

Aber Moonlets leisten noch viel mehr: Sie schützen das iPad sogar bei Stürzen, etwa auf den Boden oder den Tisch.

Die besondere Formgebung der Moonlets erhöht die Funktionalität und Lebensdauer des iPads, gleichzeitig bietet es höchsten Komfort und ergonomische Vorteile bei der Benutzung des Tablet Computers.

As simple as it is clever: Moonlet protects the iPad, and complements its design in an elegant way. It is designed in Switzerland, and manufactured in Germany using the latest patented technology.

The Moonlets are simply popped on to the corners of this tablet computer. They raise it off the surface of the workplace, protecting it against scratches and accidental damage as a result of spilled coffee, for example.

Moreover, their non-slip material – they are made from a high-quality mixture of elastomer and rubber – make the iPad easy to use at any angle when typing on or working with it. The small rubber balls, which are available in black or white, sit excellently in the hand, and their pleasant satin surface make them very easy to grip.

But Moonlets can do much more: they can even protect the iPad if it falls over or is dropped.

The special design of the Moonlets enhances the iPad and gives it a longer life. At the same time, it makes using the tablet computer extremely comfortable.

19:08
iPad

iLam
Heißlaminiergerät

Hot roller laminator

Kommunikation,
Audio / Video, Optik
Silber

Communication,
audio / video, optical equipment
Silver

Jury
Judges panel

Neue Heißlaminiergeräte mit interessanten Innovationen. Vollautomatisch und selbsterklärend. Es wurde in einem schwierigen Marktsegment eine eigenständige neue Struktur gefunden, die sich auch auf verschiedene Größen übertragen lässt. Im Gegensatz zu den üblichen Geräten sehr kompakt, zusammenklappbar und platzsparend, was besonders für die temporäre Benutzung wichtig ist.

New hot roller laminator with interesting innovations. Fully automatic and self-explanatory. In a difficult market segment, a distinctive new structure has been found, which can also be scaled to different formats. Unlike the usual appliances, very compact. Can be folded up again to save space. This is especially important for one-off use.

Hersteller
Manufacturer
Esselte Leitz GmbH & Co. KG
D-70469 Stuttgart

Design
Designer
Weinberg + Ruf
D-70794 Filderstadt

Vertrieb
Distributor
Esselte Leitz GmbH & Co. KG
D-70469 Stuttgart

Laminiert wird in vielen Bereichen: von Fotos und Rezepten über Namensschilder, Sicherheitshinweise und Bedienungsanleitungen bis hin zu Menükarten und Hotelinformationen. Die meisten Laminiergeräte sind kompliziert in der Bedienung, und es müssen viele Einstellungen, wie Folienstärke, Temperatur, Geschwindigkeit et cetera, vorgenommen werden.

Dem macht nun Leitz ein Ende und hat Laminiergeräte entwickelt, die extrem einfach zu bedienen sind. Dank einer technischen Innovation sind nun keine Einstellungen mehr erforderlich. Die Geräte arbeiten vollautomatisch, erkennen selbst die Dokumentendichte und stellen die entsprechende Geschwindigkeit ein mit perfekten Laminierergebnissen – und alles bei niedrigem Energieverbrauch. Die Handhabung ist selbsterklärend, die Bedienungsanleitung muss nicht mehr vorher gelesen werden – ideal auch für Mehrfachnutzung in Büros oder Schulen.

Zu dem Gerätesortiment gehören zwei Produktlinien: iLam touch für den Bürobereich und iLam easy für Büro und privates Umfeld, jeweils in einer A4- und einer A3-Ausführung. Die kompakten Geräte machen einen robusten und hochwertigen Eindruck.

Die patentierten, einklappbaren Ein- und Ausgabefächer bei iLam touch sorgen für glatte Ergebnisse. Bei iLam easy wurde die Ausgabe abgeschrägt, um zu verhindern, dass sich das noch warme Laminat verbiegt.

There are many things that can be laminated: photos and recipes, name tags, safety notices and operating instructions, menus and notes for hotel guests. Most laminators are complicated to use, with lots of settings having to be made, such as sheet thickness, temperature, speed, and so on.

Leitz has now put an end to this, developing a laminator that is extremely easy to use. Thanks to a technical innovation, no more settings are needed. The laminator works fully automatically, detecting document thickness on its own and setting the required speed. The result is perfectly laminated documents – and this while consuming little energy. Operation is self-explanatory, with no need to read to operating instructions beforehand – which is ideal when it is being used by many different people in an office or school.

The range includes two product lines: iLam touch for the office and iLam easy for offices and private use. Each is available in an A4 and A3 model. The compact appliances give a sturdy, high-quality impression.

In the iLam touch, the patented fold-in input and output flaps ensure smooth results. In the iLam easy, the output tray has been bevelled to prevent the hot laminate from bending.

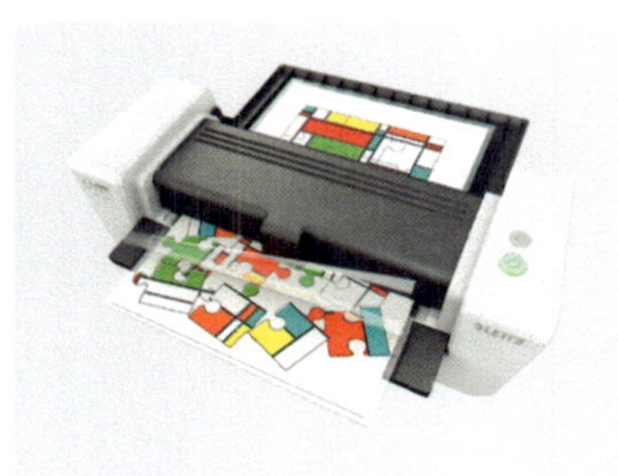

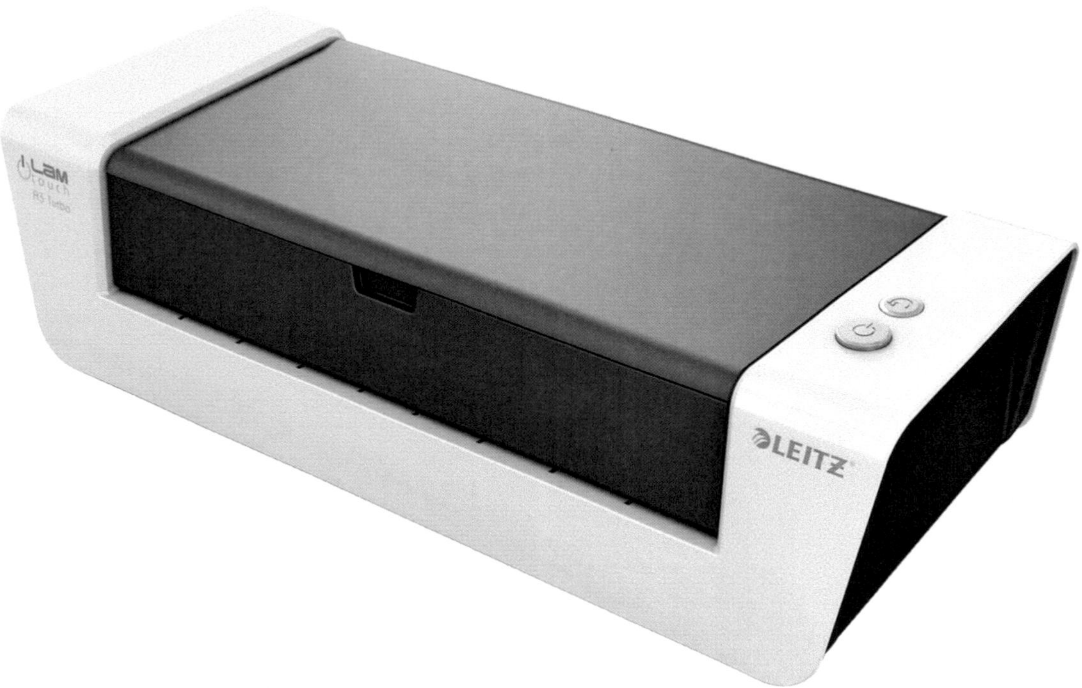
LEITZ

Jury
Judges panel

World of Wagner
Imagefilm

**Kommunikation,
Audio / Video, Optik
Silber**

Image film

**Communication,
audio / video, optical equipment
Silver**

Ein Imagefilm, der Design ganz anders zum Thema macht, als man es von einer Firma mit diesem Produkthintergrund erwartet. Das finden wir sehr schön gemacht. Die Qualität der Ausführung würde man eigentlich nur von sehr viel teureren oder hochwertigeren Produkten erwarten, zum Beispiel wenn Porsche diese Art Imagefilm mit so einem professionellen Sprecher hätte, und nicht von fast banalen Produkten wie Türstopper, Möbelgleiter oder Räder. Kaum ein Film vermittelt so viel Optimismus und positive Energie wie dieser Film.

An image film that treats its topic, design, much differently than you would expect from a company that makes the sort of products it does. We think it has been very well done. The quality of its execution is one you would really only expect from much more expensive or prestigious products – the sort of image film you would expect from Porsche, with such a professional speaker. It's not the sort of film you would expect for such banal products as door stoppers, furniture glides or casters. Hardly any film conveys so much optimism and positive energy as this one.

Moving pictures can be just that – incredibly moving. They can say more than a thousand words. The new high-definition image film »World of Wagner« is great cinema in just three minutes. When a father and son work together to a build a »dream car« – a soapbox car – a heavenly light floods the room and the child's eyes light up with wonder.

Or a sprightly old lady effortlessly manoeuvres an oversized house plant on casters. When new artworks are delivered to a gallery. Or a young couple move in together for the first time. When book shelves tower up from nothing to dizzying heights, the latest digital technology is at work.

All this is Wagner. Or rather: all this is people at home or at work, where Wagner products – casters, transport aids, door stoppers, furniture glides, furniture legs – play a role.

Hersteller
Manufacturer
Wagner System
D-77933 Lahr

Design
Designer
Bar Vinya Film
Jan Reiff
D-77797 Ohlsbach

Vertrieb
Distributor
Wagner System
D-77933 Lahr

Bewegte Bilder können unglaublich bewegend sein und mehr sagen als 1000 Worte. So zeigt der neue, in HD-Qualität produzierte Imagefilm »World of Wagner« in drei Minuten großes Kino: Wenn der Vater mit dem Sohn gemeinsam am »Traumwagen« – der Seifenkiste – bastelt, bricht überirdisches Licht in den Raum und Kinderaugen staunen und leuchten.

Oder eine rüstige Dame manövriert eine überdimensional groß gewachsene Grünpflanze spielend auf Rollen. Wenn in einer Galerie neue Kunst angeliefert wird. Oder ein junges Paar zum ersten Mal zusammenzieht. Wenn sich Bücherregale aus dem Nichts in schwindelerregende Höhen bauen, ist das neueste Digitaltechnik.

Das alles ist Wagner. Oder anders. Das alles sind Menschen in ihrem Zuhause oder Arbeitsbereich, in dem Wagner-Produkte – wie Räder, Transporthilfen, Türstopper, Möbelgleiter, Möbelbeine – eine Rolle spielen.

INTER TRANS
UMZÜGE

Hersteller
Manufacturer
Supernova
D-79194 Gundelfingen

Design
Designer
Werksdesign
In-house design
Marcus Wallmeyer

Vertrieb
Distributor
Supernova
D-79194 Gundelfingen

INFINITY 8 SUPER-NOVA

Infinity 8
Ausschaltbare Fahrrad-dynamonabe

Switchable dynamo hub

Freizeit, Outdoor
Gold

Leisure, outdoor
Gold

Infinity 8
Ausschaltbare Fahrrad-dynamonabe

Switchable dynamo hub

Jury
Judges panel

Freizeit, Outdoor
Gold

Leisure, outdoor
Gold

Bietet traditionelle Funktionen mit einer innovativen Technologie an: Der Rollwiderstand im Nabendynamo ist abschaltbar. Bei Fahrrädern mit herkömmlichen Nabendynamos wird man gehemmt und braucht mehr Kraft, um Licht zu erhalten. Das ist hier nicht der Fall. Ein weiterer Vorteil: Man hat das Licht immer dabei. Ein einfaches Produkt, sehr überzeugend, sehr ästhetisch.

Offers traditional functions using innovative technology. The drag from the hub dynamo can be eliminated. On bicycles with conventional hub dynamos, there is always resistance, meaning the rider needs more power to generate a light. This is not the case here. And there is a further plus point: the light is always available. A simple product, very impressive, very tasteful.

Hersteller
Manufacturer
Supernova
D-79194 Gundelfingen

Design
Designer
Werksdesign
In-house design
Marcus Wallmeyer

Vertrieb
Distributor
Supernova
D-79194 Gundelfingen

Infinity 8 ist ein Meisterwerk der Ingenieurskunst. In dem Nabendynamo wurden gleich zwei neue revolutionäre, patentierte Konzepte verwirklicht.

Das eine ist die raffinierte FreeSpin-Technologie, durch die Infinity 8 zu 100 Prozent abgeschaltet werden kann. Das macht ihn dem normalen Nabendynamo haushoch überlegen. Einfach am Schaltmechanismus drehen und den unerwünschten Leerlaufwiderstand ausschalten. Das spart Kraft und macht aus dem Dynamo im Handumdrehen eine leichtlaufende Vorderradnabe. Die Leistung wird nur abgerufen, wenn man sie wirklich benötigt.

Die zweite bahnbrechende Innovation ist die zukunftsweisende SoloMagnet-Technologie. Herkömmliche Dynamos mit Magneten auf der Innenwand des Nabengehäuses können nur einen Pol zur Stromerzeugung nutzen. Der Infinity 8 dagegen besitzt einen zentral angeordneten Hochleistungsmagneten, der beide Magnetpole effektiv einsetzt. Damit liefert er bereits bei geringen Geschwindigkeiten genügend Leistung für optimales Licht.

Zudem steht mit dem Ladeadapter »The Plug« selbst bei langsamer Fahrt Strom für GPS, Handy oder zum Nachladen der Airstream Akkulampe zur Verfügung.

Infinity 8 is an engineering masterpiece. The dynamo hub incorporates two revolutionary new concepts, which have been patented.

The first is its sophisticated FreeSpin technology, which allow Infinity 8 to be completely switched off. This makes it far superior to the normal dynamo hub. A simple turn of the dial-wheel mechanism is enough to eliminate unwanted drag when the dynamo is running idle. This saves pedalling power, and turns the dynamo into a smooth-running front wheel hub. It only provides power when it is really needed.

The second ground-breaking innovation is its trendsetting SoloMagnet technology. Conventional dynamos with magnets attached to the walls of the hub body use only one magnet pole to generate electricity. Infinity 8, by contrast, has a high-power magnet at its core. This means it can effectively use both poles of the magnet. As a result, it delivers enough power for optimum lighting even at low speeds.

In addition, its »The Plug« USB power supply allows a GPS to be powered, or a mobile phone or Airstream battery light to be recharged even at low speeds.

OFF
on

SUPERNOVA
SUPERNOVA
AIRSTREAM

Airstream
Wiederaufladbare LED-Fahrradlampe

Rechargeable LED bicycle lamp

Freizeit, Outdoor
Silber

Leisure, outdoor
Silver

Jury
Judges panel

Sehr aktuelle LED-Leuchte, die extrem hell, einfach zu bedienen und formal schön ist. Sie kann über ein Netzteil, aber auch über USB und Dynamo geladen werden. Die Ladestandsanzeige wird visualisiert durch ein zusätzliches Lichtelement. Ein sehr hochwertig gemachtes Produkt. Auch die Wärmeabstrahlung ist ästhetisch arrangiert. Die Leuchte entwickelt sich im Grunde aus dem Kühlkörpermodul für die LED. Schnell an- und abzubauen, auch als Taschenlampe gut zu verwenden.

Very modern LED lamp: extremely light, simple to operate, and formally elegant. It can be recharged using a power supply unit or a USB, and from a dynamo. An additional light element visualizes the level of battery charge. A product made to very high quality standards. The cooling system has also been designed tastefully. The lamp has basically evolved out of the cooling module for the LED. Quick to mount and detach, is also suitable as a pocket torch.

Hersteller
Manufacturer
Supernova
D-79194 Gundelfingen

Design
Designer
Werksdesign
In-house design
Marcus Wallmeyer

Vertrieb
Distributor
Supernova
D-79194 Gundelfingen

Die vielseitig einsetzbare Fahrradleuchte ist mit einer maximalen Helligkeit von bis zu 370 Lumen extrem hell. Sie breitet einen weitreichenden Lichtteppich vor dem Fahrer aus.

Ob auf dem Weg zur Arbeit, im dichten Stadtverkehr, beim Rennradtraining oder mit dem Mountainbike auf schnellen Trails – die Airstream ist immer ein idealer Begleiter. Selbst bei gedimmtem Licht – mit Leuchtzeiten von bis zu sieben Stunden – ist sie hell genug, um damit Rad zu fahren. Sie erfüllt auch die strengen Vorschriften der deutschen Straßenverkehrszulassung.

Untergebracht ist sie in einem kompakten, robusten und wassergeschützten Gehäuse aus Aluminium. Sie besitzt eine wegweisende Hitzeregulierung, die es möglich macht, die LED und den integrierten und hochwertigen Lithium-Ionen-Akku bei optimalen Temperaturen zu betreiben. Dadurch bleiben die vortrefflichen Leistungsdaten dauerhaft erhalten. Eine rote und eine grüne LED neben der beleuchteten Bedientaste zeigen die Akkukapazität an. Sie kann über Netzteil, USB und selbst über Dynamo geladen werden. Damit ist sie ideal für Langstreckenfahrer.

Mittels ihrer universellen Schnellbefestigung lässt sie sich ganz leicht am Fahrrad anbringen. Da der Lenkerhalter demontierbar ist, kann man die Fahrradleuchte auch als extrem kraftvolle Taschenlampe nutzen.

This versatile bicycle lamp is extremely bright. It generates a maximum brightness of 370 lumen, and creates a broad and far-reaching illuminated area to the front of the rider.

Whether on the way to work, in heavy city traffic, while road training or mountain biking on fast trails, the Airstream is always an ideal companion. Even when the light is dimmed – when one battery charge will last up to seven hours – it is still bright enough to see by. It also fulfils the strict German road safety regulations.

Its compact, sturdy and waterproofed housing is made of aluminium. It is fitted with a pioneering cooling system that allows the LED and the integrated, high-quality lithium-ion rechargeable battery to be operated at optimum temperature. This means that its excellent performance is guaranteed at all times. A red and a green LED next to the illuminated on/off switch show the level of charge. It can be recharged using a power supply unit, a USB, and even from a dynamo. This makes it ideal for long-distance riders.

Its universal mount allows it to be attached easily to the bicycle. As the handlebar mount is detachable, the bicycle lamp can also be used as an extremely powerful pocket torch.

Electric Fire
E-Bike / Pedelec

E-bike / Pedelec

Freizeit, Outdoor
Silber

Leisure, outdoor
Silver

Jury
Judges panel

Wir finden, dass dieses E-Bike ein sehr sportliches Fahrrad ist. Wobei das Konzept nicht einfach in ein Mountain-Bike gepresst ist. Es ist ein E-Bike, das exzellente Eigenschaften und durch die Bosch-Elektronik eine gute Reichweite hat, über 50 Kilometer, wenn man ökonomisch fährt. Es wirkt nicht nur von der Optik her leicht, sondern fährt sich auch leicht und sehr gut. Außerdem wiegt es für ein E-Bike mit 21 Kilo wenig. Einfach schön.

For an e-bike, we feel this is a very sporty bike. Yet the idea has not simply been forced into a mountain bike design. It's an e-bike with excellent properties and, thanks to the Bosch electronics, a good range: more than 50 kilometres if is ridden economically. It not only looks light, but also offers a smooth, excellent ride. And at 21 kilos, it doesn't weigh much for an e-bike. Simply appealing.

Hersteller
Manufacturer
Merida & Centurion
Germany GmbH
D-71106 Magstadt

Design
Designer
Slogdesign
Jürgen Hinderhofer
D-88400 Biberach

Vertrieb
Distributor
Merida & Centurion
Germany GmbH
D-71106 Magstadt

Im Blickpunkt der Fortbewegung auf zwei Rädern stehen mittlerweile E-Bikes, denn sie eröffnen vielen Menschen neue Möglichkeiten, das Fahrrad im Alltag, zum Einkaufen oder als Hobby zu nutzen. Gemeinsam mit Bosch wurde ein sportliches Strom-Rad, das in vier Versionen erhältlich ist, umgesetzt.

Herzstück des Electric Fire ist sein Rahmenkonzept. Durch das leicht zum Sitzrohr hin abfallende Oberrohr tritt das Bike bewusst sportlich auf. Die komfortable Forke, die 30 Millimeter Federweg zur Verfügung stellt, unterstreicht diesen Anspruch. Das Oberrohr läuft zweigeteilt vom Steuer- zum Sitzrohr und bietet innerhalb einer separaten Abdeckung genügend praktischen Stauraum für Pumpe, Flickzeug, Haustürschlüssel und dergleichen mehr.

Ebenso sportlich wie der optische Auftritt ist sein kraftvoller Antrieb. Er ist als Mittelmotor-Konzept konstruiert. Der 250-Watt-Motor sitzt unterhalb des Tretlagers, was dem Electric Fire einen tiefen Schwerpunkt verschafft. Dadurch ist das Bike ungemein agil und wendig. Dank seinem Gesamtgewicht von 21 Kilo und seinen nominal 50 Newtonmetern Drehmoment beschleunigt es ausgesprochen kraftvoll. Dabei analysiert die ausgeklügelte Bosch-Elektronik permanent den Fahrerwunsch nach Vortrieb und setzt die insgesamt zwölf möglichen Fahrmodi jederzeit höchst effizient um.

Ein weiterer Vorteil der Drei-Sensoren-Elektronik: Anfahren und Abbremsen gelingen mit dem Electric Fire absolut ruckelfrei, sämtliche Geschwindigkeitswechsel während der Fahrt ebenso.

When it comes to mobility on two wheels, the focus of attention is now on e-bikes. For many people, they open up new possibilities of using a bicycle in everyday life, when shopping, or as a hobby. Together with Bosch, a sporty electric bike has been developed that is available in four versions.

The heart of Electric Fire is its frame design. With its crossbar sloping slightly down toward the seat post, the bike is deliberately sporty in appearance. The front fork, whose 30-millimetre spring travel provides riding comfort, underscores this image. The crossbar between the steering column and seat post is in fact two crossbars, and underneath the cover between them there is plenty of practical stowage space for a pump, repair kit, door key and the like.

But it is not only its image that is sporty. Its drive unit is also powerful. It has been designed as a central motor. Delivering 250 watts, it is mounted beneath the bottom bracket, giving Electric Fire a low centre of gravity. This makes the bike unusually agile and manoeuvrable. Thanks to its total weight of 21 kilos and its torque of a nominal 50 Newton meters, it accelerates very powerfully. The sophisticated Bosch electronics permanently analyzes the rider's propulsion needs, and translates them highly efficiently into 12 possible riding modes.

The three-sensor electronics is a further plus point. Riding off and braking with Electric Fire is absolutely jolt-free, as are all speed changes while riding.

CENTURION

Giro Vitale
Radtrainer

Pedal trainer

Jury
Judges panel

Freizeit, Outdoor
Silber

Leisure, outdoor
Silver

Eine sehr überzeugende nutzerorientierte Entwicklung, weil ältere Menschen, die nicht mehr so mobil sind, damit trainieren und dabei ihre Mobilität fördern können, während sie auf der Bank sitzen und sich miteinander unterhalten. Wir finden das Konzept prima.

A very convincing, user-focused development, because it allows older people who are no longer so agile to do some exercise and thus improve their agility while they sit on a bench and chat with each other. We think this is a super idea.

Hersteller
Manufacturer
L. Michow & Sohn GmbH
D-22041 Hamburg

Design
Designer
Grauwert
Büro für demografiefeste Produkte und Dienstleistungen
Mathias Knigge
D-22761 Hamburg

Vertrieb
Distributor
L. Michow & Sohn GmbH
D-22041 Hamburg

Eine Voraussetzung, sich auch im Alter wohlzufühlen, ist, beweglich zu bleiben. Das unterstützt das innovative Konzept Giro Vitale. Es bietet die Möglichkeit, sich an frischer Luft an speziell entwickelten Geräten zu bewegen.

Dabei werden Muskulatur, Kondition, Gleichgewicht und Koordination trainiert. Die Geräte animieren nicht zu Höchstleistungen, sondern laden zu gezielten, einfachen Übungen ein, die Spaß an Bewegung vermitteln und das geistige und körperliche Wohlbefinden sowie die Mobilität der älteren Menschen fördern.

Eines dieser Geräte ist der Radtrainer. Mit ihm können auf sanfte Art Bein- und Rückenmuskulatur trainiert werden, aber in bequemer Sitzhaltung und ohne dass man auf ein angeregtes Gespräch mit seinem Sitznachbarn verzichten muss. Der moderate Widerstand erfordert dabei wenig Kraft.

Die dazugehörige Ruhebank wurde speziell für Giro Vitale entwickelt. Sitzhöhe und Position sind auf die Bedürfnisse älterer Benutzer abgestimmt. Ein weiteres Produktmerkmal bietet einen zusätzlichen Nutzen: Die Bank besitzt Schlitze, in denen gehbeeinträchtigte Menschen ihre Gehhilfen abstellen können.

Neben dem Radtrainer gehören zu dem Konzept noch weitere Bewegungselemente wie Rundlauf, Pendelscheibe oder Nordic Trainer. Sie alle zeichnen sich durch eine klare Gestaltung aus und sind für den Einsatz im öffentlichen Raum und in Parks ebenso wie für den halböffentlichen Bereich, wie etwa Wohnanlagen und Reha-Kliniken, konzipiert.

One condition for well-being in old age is remaining nimble. The innovative Giro Vitale concept helps to achieve this. It offers the possibility of exercise in the fresh air, using specially designed apparatus.

This trains musculature, fitness, balance and coordination. The apparatus is not about high performance, but invites users to do specific, simple exercises that make movement a pleasure and promote mental and physical well-being, as well as mobility, among the elderly.

One of these pieces of apparatus is the pedal trainer. It can be used to gently train the leg and back muscles from a comfortable seating position. And it doesn't stop users from having an animated conversation with their neighbours. The moderate resistance does not require a lot of strength.

The bench that goes with it was developed especially for Giro Vitale. The seating height and position were designed with elderly users in mind. A further product property offers an additional benefit. The bench is furnished with slits into which people with impaired mobility can place their walking sticks.

Apart from the pedal trainer, the range includes further exercise apparatus such as a treadmill, balance trainer, and Nordic trainer. All of them are uncluttered in design, and ideal for use in public spaces and parks, as well as in semi-public spaces such as old people's homes and rehabilitation clinics.

Twitter
Nistkasten

Nesting box

Jury
Judges panel

Freizeit, Outdoor
Silber

Leisure, outdoor
Silver

Minimalistisch. Es ist das, was man haben möchte: einen Vogel im Garten und ein ästhetisch wertvolles Objekt am Baum. Man hat verschiedene farbliche Variationen und zwei einfache Materialien. Ein gutes Beispiel für Minimal Art mit noch einem zusätzlichen praktischen Feature durch das Edelstahldach. Das kleinste Designhotel der Welt.

Minimalistic. It gives you what you want – a bird in your garden – plus an aesthetically valuable object attached to your tree. There are various colour combinations. Simple materials, just two of them. A good example of minimal art, with an additional practical feature in the form of a stainless-steel roof. The world's smallest designer hotel.

Hersteller
Manufacturer
Keilbach design
and products
D-74677 Dörzbach

Design
Designer
Werksdesign
In-house design
Peter Keilbach

Vertrieb
Distributor
Keilbach design
and products
D-74677 Dörzbach

Der würfelförmige Nistkasten wirkt wie ein Designobjekt und setzt selbst im kleinsten Garten ungewöhnliche Akzente. Er besticht durch seine kubische und auf das Wesentliche reduzierte Form.

Die äußere Hülle besteht entweder aus Edelstahl oder Cortenstahl, der als Regendach dient. Das Innere dagegen ist komplett aus Holz und bietet damit Vögeln einen artgerechten Nistplatz. Die Stahlabdeckung liegt nicht direkt auf dem hölzernen Innenraum an, sodass die Luft dazwischen zirkulieren kann.

Der ungewöhnliche Nistkasten kann an einen Ast gehängt oder an einem Baumstamm befestigt werden. Steht kein Baum zur Verfügung, lässt er sich auf einen Pfahl schrauben oder an der Hauswand anbringen und wird so zum unverwechselbaren bewohnbaren Designobjekt.

Twitter ist in zahlreichen Farb- und Materialvarianten erhältlich: schlicht in unbehandeltem Holz mit Corten- oder Edelstahl oder die Edelstahlversion mit lackiertem Holz in den Farben Gelb, Orange, Pink, Blau und Braun. Bei den lackierten Ausführungen handelt es sich um eine wetterfeste umweltfreundliche Farbe.

Damit bietet Twitter nicht nur nur einen ästhetischen Blickfang, sondern ist auch das kleinste Designhotel der Welt für architektur- und designbegeisterte Vögel.

This cube-shaped nesting box looks like a designer object, and sets an extraordinary accent in even the smallest of gardens. Its cubist form, reduced to the bare essentials, makes it stand out.

Its outer skin is made of either stainless steel or corten steel, which keeps off the rain. On the inside, by contrast, it is made completely of wood, offering birds an appropriate nesting place. A gap between the steel cover and the wooden interior allows the air to circulate.

This unusual nesting box can be suspended from a branch or attached to a tree trunk. If there is no tree available, it can be screwed to a post or attached to the wall of a house, creating an unmistakable designer-object home.

It is available in many colour and material combinations: there is a plain version with untreated wood in combination with corten or stainless steel, or the stainless steel version with wood painted yellow, orange, pink, blue or brown. The paint used for the latter variants is weatherproof and eco-friendly.

Twitter is thus not only a tasteful eye-catcher, but also the world's smallest designer hotel for feathered design and architecture enthusiasts.

Nimbletoes® gripster
Schuhe

Shoes

Jury
Judges panel

Freizeit, Outdoor
Silber

Leisure, outdoor
Silver

Ein extrem bequemer Schuh, in dem man sich fühlt, als würde man barfuß laufen. Er sitzt hervorragend am Fuß und hat eine besondere Schnürung mit einem umlaufenden Band, das die Sohle an den Fuß bindet. Dies unterstützt die Fußstabilität in diesem Schuh ausgezeichnet.

This is an extremely comfortable shoe, in which you feel as though you were walking barefoot. Fits the foot excellently. It has a special lacing with laces going right around, binding the sole to the foot. This gives excellent support to walking stability in this shoe.

Hersteller
Manufacturer
Bär GmbH
D-74321 Bietigheim-Bissingen

Design
Designer

Schaft
Upper
Werksdesign
In-house design
Michael Bach
und
Solveig Gubser
D-10623 Berlin

Sohle
Sole
Rucky Zambrano
CH-6864 Arzo

Vertrieb
Distributor
Bär GmbH
D-74321 Bietigheim-Bissingen

So viel Schuh wie nötig, so viel barfuß wie möglich – lautet das Motto der Nimbletoes®. Durch die besondere Passform und Sohle kommt das Gehen in diesen Schuhen dem Barfußlaufen sehr nahe. Das stärkt und kräftigt die Fußmuskulatur sowie die Bänder und Sehnen. Zugleich kann das natürliche Abrollen die Knie- und Hüftgelenke stark entlasten.

Gripster ist ein Multi-Sportschuh, der für Fitness, Yoga, Pilates und alles, bei dem Dynamik im Spiel ist, seinen Einsatz findet. Er besitzt einerseits Grip – er ist standhaft und rutschfest –, andererseits schmiegt er sich wie ein Handschuh an den Fuß. Zudem ist er bodensensitiv.

Dazu kommt die innovative Archraiser®-Schnürung, die dem Fußbogen optimalen Halt gibt. Diese greift unter dem Fuß hindurch und fasst alle Knochen, Muskeln und Sehnen des Mittelfußes zu einer kraftvollen, federnden Einheit zusammen. Die Fußbögen erhalten so eine ganz neue Elastizität und fühlen sich an, als würde man sie von oben anheben. Die Füße müssen sich nicht mehr an ein Fußbett gewöhnen, das nie an sie angepasst wurde.

Gripster besitzt einen atmungsaktiven Textilschaft, eine Vibram®-Gummisohle und im Innern eine zwei Millimeter dünne XRD-Dämpfung, die Stöße optimal abfedert. Ein weiteres innovatives Produktmerkmal ist die Zehenfreiheit®, die der Schuh bietet. Ohne eingeengt zu sein, können sich die Zehen nun nicht nur ungehindert spreizen, sondern auch natürlich abrollen.

As much shoe as necessary, as much barefoot as possible – this is the motto of Nimbletoes® shoes. With their special fit and sole, walking in these shoes comes close to walking barefoot. This strengthens the musculature of the feet, and well as the tendons and ligaments. At the same time, the natural rolling of the foot takes a considerable burden away from the knee and hip joints.

gripster is a shoe for all kinds of sport. It can be used for workouts, yoga, pilates and wherever energetic movement is needed. On the one hand, it provides grip – it is firm and non-slip – and on the other it fits the foot like a glove. And because the wearer can sense the ground, it is like walking barefoot.

It also features innovative Archraiser® lacing, which gives the arch of the foot optimum support. It reaches underneath the foot and brings together all the bones, muscles and tendons of the metatarsus to form a powerful, elastic unity. In this way, the arches are given a completely new elasticity, and feel as though they were being raised from above. The feet no longer have to get used to a footbed that was never moulded to fit them.

gripster has a breathable textile upper, a Vibram® rubber sole and a two-millimetre-thick XRD shock-absorbing layer on the inside. One further innovative product feature is the freedom of movement the toes have inside the shoe. Without being cramped, the toes can now not only spread freely, but also roll naturally.

FUN
FUN
FACTORY

Yooo
Erotic Toy Multizone Vibrator

Freizeit, Outdoor
Silber

Erotic toy multizone vibrator

Leisure, outdoor
Silver

Jury
Judges panel

Multifunktional einsetzbar sowohl für die Dame im Intimbereich als auch für den Herrn auf Schulter, Brust et cetera. Es ist ein Gerät, das man auch im Badezimmer liegen lassen kann, damit es andere begutachten können. In anderen Kulturen kann es als Massagegerät verkauft und damit unauffälliger erworben werden. Es hat eine sehr smarte Aufladefunktion, einfach den Adapter aufs Logo, auf Fun, setzen.

Multifunctional use, both for women in the genital area and for men on the shoulder, under the armpits, on the chest, and so on. It's a device I would also leave lying around in the bathroom so that others can inspect it. In other cultures, it can be sold as a massage device, and thus be purchased more discretely. The device has a smart recharging function: simply set the adapter to the logo, to Fun.

Hersteller
Manufacturer
Fun Factory GmbH
D-28197 Bremen

Design
Designer
Werksdesign
In-house design
Nina Smerzinski

Vertrieb
Distributor
Fun Factory GmbH
D-28197 Bremen

Der Multizone Vibrator besteht aus drei Kugeln in zwei unterschiedlichen Größen und lässt sich damit individuell nach Lust und Laune einsetzen. Durch seine völlig neue, reduzierte Form und die sinnliche Ästhetik lädt er zum Experimentieren ein, dank der innovativen Click 'n' Charge-Aufladetechnologie, ohne störende Unterbrechungen.

Yooo ist vielseitig einsetzbar – nicht nur im Intimbereich, sondern auch bei Verspannungen im Nacken und der Rückenmuskulatur.

Die Besonderheit des Erotic Toys liegt in seiner Vielseitigkeit. So ist es mit zwei Motoren ausgestattet, die langlebig, stark und doch leise sind. Mit ihnen werden asynchrone Vibrationsmuster erzeugt, die eine Vielzahl an intensiven, facettenreichen Stimulationen und spannenden Vibrationseffekten ermöglichen.

Yooo ist bequem zu handhaben, einfach und intuitiv über Tipptasten zu bedienen und stufenlos regulierbar. Das Bedienelement mit der Magic-Touch-Funktion leuchtet auf, wenn es berührt wird, und zeigt damit an, dass der Vibrator einsatzbereit ist.

Der Vibrationskörper besteht aus hundert Prozent hochwertigem Silikon, das sehr strapazierfähig, anti-allergen und leicht zu reinigen ist. Mit dem dazugehörigen Aufladestecker »Fun Magnetic Plug« lässt sich Yooo einfach an der Steckdose aufladen.

This multizone vibrator comprises three balls of two different sizes, which means it can be used to suit individual mood and desire. With its completely new, reduced form and sensual aesthetic, it invites the user to experiment. Thanks to its innovative click 'n' charge recharging technology, the fun is uninterrupted.

Yooo is versatile – apart from the genital area, it can also be used to combat back and neck tension.

Its versatility is what makes this erotic toy special. It is equipped with two motors, which are long-lasting, powerful and yet quiet. They are used to generate asynchronous vibration patterns that allow all kinds of intensive and multifaceted simulation and exciting vibration effects.

Yooo is pleasant to hold, simple and intuitive to operate using jog keys, and infinitely adjustable. The operating buttons with their magic touch function light up when they are touched, showing that the vibrator is ready for use.

The vibrator body is made completely of high-quality silicone, which is very hard wearing, anti-allergenic, and easy to clean. Using the »Fun Magnetic Plug«, Yooo can be recharged simply from the electricity main.

G4 Vibes
Vibratorserie

Series of vibrators

Freizeit, Outdoor
Silber

Leisure, outdoor
Silver

Jury
Judges panel

Formal schöner Vibrator in mehreren Ausführungen. Einfach zu bedienen und selbsterklärend. Reagiert touch-sensitive. Gute Tastenbedienung. Endlich mal ein Vibrator, der keine übliche Fleischdarstellung und nicht überbordend oder niedlich ist, sondern richtig gut gestaltet wurde. Überzeugend auch die Durchgreifmöglichkeit am Bedienelement. Er ist leise und schaltet sich nicht aus Versehen an, somit ist er auch mitreisefähig. Eine extrem smarte Aufladefunktion, einfach den Adapter aufs Logo, auf Fun, setzen.

Formally elegant vibrator in several versions. Simple to use and self-explanatory. Reacts touch-sensitively. Good key operation. At last a vibrator without the usual flesh connotations. Nor is it over the top or twee, just really well designed. The way you can put your finger through the operating element is also good. It is silent, and won't switch on accidentally, making it something you can put in your suitcase. An extremely smart recharging function: simply set the adapter to the logo, to Fun.

Hersteller
Manufacturer
Fun Factory GmbH
D-28197 Bremen

Design
Designer
Werksdesign
In-house design
Verena Michel
und
Phoenix Design GmbH & Co. KG
D-70376 Stuttgart

Vertrieb
Distributor
Fun Factory GmbH
D-28197 Bremen

Die Vibratorserie G4 Vibes setzt völlig neue Maßstäbe in Bedienung und Funktionalität. Die innovative Gestaltung des Loop-Bedienelementes und die Magic-Touch-Beleuchtung sind nicht nur einfach im Handling und intuitiv in der Bedienung, sondern auch elegante Hingucker. Das Bedienelement ist ergonomisch geformt und besteht aus hochglänzendem, hypoallergenem Kunststoff mit einem ungewöhnlichen Metalleffekt im Inneren des Loops.

Über Tipptasten lassen sich die modern und ästhetisch gestalteten Vibratoren stufenlos regeln. Dank der Magic-Touch-Funktion leuchtet das Bedienteil, wenn es berührt wird, und zeigt an, dass der Vibrator einsatzbereit ist.

Die Vibrationskörper bestehen aus hundert Prozent medizinischem Silikon. Dadurch sind sie angenehm flexibel und fühlen sich besonders samtig an. Außerdem ist das Material sehr strapazierfähig, anti-allergen und leicht zu reinigen.

Die Vibratorserie G4-Vibes überzeugt aber auch durch die innovative Click 'n' Charge-Aufladetechnologie. Mit dem dazugehörigen Aufladestecker »Fun Magnetic Plug« lassen sich die Vibratoren einfach an der Steckdose aufladen.

Zu der G4-Vibes-Serie gehören die formschönen Vibratoren Elegance, Patchy Paul, Tiger, Calla sowie Paul & Paulina. Sie sind alle in verschiedenen frischen und kräftigen Farben erhältlich.

The G4 Vibes vibrator series sets completely new standards when it comes to operation and functionality. The innovative design of the loop operating element and the magic touch lighting are not only easy to handle and intuitive to use, but also elegant eye-catchers. The operating element is ergonomically shaped, and made of highly polished, hypo-allergenic plastic with an unusual metal effect on the inside of the loop.

Using jog keys, the intensity of the modern and tastefully designed vibrators can be infinitely varied. Thanks to the magic touch function, the operating buttons light up when they are touched, showing that the vibrator is ready for use.

The body of the vibrator is made of 100 percent medical silicone. This makes them pleasantly flexible and especially velvety to the touch. In addition, the material is very hardwearing, anti-allergenic and easy to clean.

A further convincing feature of the G4 Vibes vibrator series is its innovative click 'n' charge recharging technology. Using the »Fun Magnetic Plug«, the vibrators can be recharged simply from the electricity main.

The G4 Vibes series comprises the elegant vibrators Elegance, Patchy Paul, Tiger, Calla and Paul & Paulina. All are available in various fresh, strong colours.

Hersteller
Manufacturer
Vaude Sport GmbH & Co. KG
D-88069 Tettnang

Design
Designer
Werksdesign
In-house design
Hans-Thomas Langowski

Vertrieb
Distributor
Vaude Sport GmbH & Co. KG
D-88069 Tettnang

DAYTOUR 24 VAUDE

Daytour 24	
Ski- und Wintertourenrucksack	Skiing and winter trekking rucksack
Sport	Sport
Gold	Gold

Daytour 24
Ski- und Wintertourenrucksack

Skiing and winter trekking rucksack

Jury
Judges panel

Sport
Gold

Sport
Gold

Genial in seiner Funktionalität. Er hat alles, was ein Skifahrer oder Bergsteiger braucht, mit kleinen Feinheiten im Detail, Metallverschlüssen, einer Trillerpfeife direkt am Brustgurt genau da, wo sie sitzen muss. Schlaufen, die man wieder verstauen kann, praktischen Fächern. Er ist komplett durchdacht. Man merkt, dass viel Erfahrung darin steckt. Diese neuen Kombinationen überzeugen uns.

Brilliantly functional. It has everything a skier or mountaineer needs, with excellent attention to detail – metal fasteners, a whistle directly on the chest strap, exactly where it is needed. Straps that can be stowed away, practical compartments. It has been thought through from top to bottom. You can see how much experience has gone into it. These new combinations are very convincing.

Hersteller
Manufacturer
Vaude Sport GmbH & Co. KG
D-88069 Tettnang

Design
Designer
Werksdesign
In-house design
Hans-Thomas Langowski

Vertrieb
Distributor
Vaude Sport GmbH & Co. KG
D-88069 Tettnang

Der Rucksack ist der ideale Begleiter für Skitouren, beim Freeriden oder Winterbergsteigen. Dafür ist er mit allen wichtigen Features ausgestattet. Daytour 24 zeichnet sich aus durch sein reduziertes Design, seinen hervorragenden Tragekomfort und seine hohe Funktionalität.

So besitzt er eine abnehmbare Helmhalterung. Selbst mit befestigtem Helm kann auf das Hauptfach zugegriffen werden. Im Safety-Fach lassen sich Lawinensonde und -schaufel sicher verstauen, sind aber trotzdem schnell zugänglich. Die Ski werden frontal oder seitlich befestigt. Zudem besitzt der Rucksack praktische Schlaufen für Eisgeräte, wie Pickel oder Stöcke.

Die seitlichen Kompressionsgurte, der leichte, abnehmbare Hüftgurt und der Brustgurt mit Pfeife bieten zusätzlichen Komfort.

Daytour 24 ist der erste bluesign®-zertifizierte Rucksack am Outdoor Markt. Der unabhängige bluesign®-Standard für Textilverarbeitung garantiert, dass Produkte in der gesamten Herstellungskette ausschließlich mit Komponenten und Verfahren gefertigt werden, die für Mensch und Umwelt unbedenklich sind, ohne dabei Kompromisse bei Funktionalität und Design einzugehen. Das Ziel von Vaude ist es, bis 2015 das ökologischste Outdoor-Unternehmen Europas zu sein, das in seinen Produkten Umweltfreundlichkeit mit den fortschrittlichsten Technologien kombiniert. So sind auch alle textilen Anteile des Rucksacks – Mesh, Lining, Rücken- und Schulterträger – bluesign® approved fabrics ohne Einbußen bei Funktion oder Eigenschaften.

The rucksack is the ideal companion for ski-hiking, free-riding, or winter mountaineering. It is equipped with all the features needed for this. The striking features of Daytour 24 are its reduced design, excellent carrying comfort, and high level of functionality.

For example, it has a detachable helmet strap. Even when a helmet is attached to it, the main compartment is still accessible. An avalanche probe and shovel can be securely stowed in the safety compartment but can be quickly retrieved if necessary. Skis can be attached to the front or sides. In addition, the rucksack has practical straps for attaching ice axes and walking poles.

Other convenient features include the side compression straps, the light, removable hip belt, and the chest strap with a whistle.

Daytour 24 is the first bluesign®-certified rucksack in the outdoor market. The independent bluesign® standard for textiles processing guarantees that products are made throughout the entire manufacturing chain using components and processes that are harmless for people and the environment, without having to accept any compromises in functionality and design. Vaude's objective is to be Europe's most eco-friendly outdoor company by 2015, with products that combine eco-friendliness with advanced technologies. All the textile components of the rucksack – mesh, lining, back supports and shoulder straps – are made of bluesign®-approved fabrics, without any negative effects on function or properties.

VAUDE

Blue One
3 Seasons Zelt

Three seasons tent

Jury
Judges panel

Sport
Silber

Sport
Silver

Es ist aus recyceltem Polyester und aus Naturfasern hergestellt, die ungefärbt und damit ökologisch sind. Auch das Beschichtungssystem des Gestänges ist umweltfreundlich, da es auf Wasser basiert. Es gibt viel kleine, feine Details, an die gedacht wurde, wie beispielsweise, dass die beiden Türen miteinander kombiniert sind. Umweltfreundlichkeit verbunden mit hoher Funktionalität. Durch seinen Baumwollanteil und die Farbe Weiß ist es auch für südliche Länder sehr gut geeignet.

It is made of recycled polyester and natural fibres, which are undyed and thus ecological. The way the rods are coated is also ecological, since it is water-based. There are many small, nice details that have been considered, such as the way the two entrances have been combined with each other. Eco-friendliness combined with a high level of functionality. Its cotton content and white colour make it suitable for Mediterranean countries.

Hersteller
Manufacturer
Vaude Sport GmbH & Co. KG
D-88069 Tettnang

Design
Designer
Werksdesign
In-house design
Matthias Kimmerle,
Hans-Thomas Langowski

Vertrieb
Distributor
Vaude Sport GmbH & Co. KG
D-88069 Tettnang

Mit Blue One wurde mit alternativen Materialien ein umweltfreundliches Pendant zu herkömmlichen Zelten entwickelt. So bestehen Außen- und Innenzelt aus 40 Prozent organischer Baumwolle und 60 Prozent recyceltem Polyester – alles ungefärbt für maximal ökologische Verträglichkeit. Der Zeltstoff sorgt für ein besonders angenehmes Wohnklima und ist durch seinen Baumwollanteil für südliche Länder besser geeignet als herkömmliche Zelte.

Blue One ist ein Ultraleichtgewicht – es bringt lediglich 2.450 Gramm auf die Waage. Dazu trägt auch das Zeltdesign bei, das durch seine Tropfenform für ein optimales Gewichts- und Raumverhältnis sorgt.

Das Zelt besitzt ein neuartiges Gestänge, bei dem das Beschichtungsverfahren auf Wasser basiert, das nach dem Prozess einfach zur Klärung gegeben werden kann. Weniger Energieaufwand ist notwendig, im Vergleich zu herkömmlichen anodisierten Gestängen wird über 50 Prozent Energie eingespart.

Alle eingesetzten Materialien sind ungefährlich für die Umwelt und die Gesundheit. Somit ist das Zelt auch für Allergiker geeignet. Umweltfreundlichkeit verbunden mit hoher Funktionalität und Qualität. Blue One ist das erste bluesign®-Zelt der Welt. Dieser strenge Umweltstandard kontrolliert die gesamte textile Wertschöpfungskette. Damit ist Blue One ein weiterer Schritt auf dem Weg des Unternehmens zum ökologischsten Outdoor-ausrüster Europas – ein Ziel, das Vaude bis 2015 erreichen will.

In Blue One, alternative materials have been used to create an eco-friendly counterpart to conventional tents. The fabric used for the outer and inner tent is 40 percent organic cotton and 60 percent recycled polyester – all of it undyed for the minimum of environmental impact. This material creates an especially pleasant atmosphere inside the tent. The cotton used means it is more suitable for Mediterranean countries than conventional tents.

Blue One is an ultra-light tent, weighing in at just 2,450 grams. This low weight is also a result of the tent design, whose tear-drop shape ensures a optimum weight-to-space ratio.

The tent has an innovative frame, which has been coated using a water-based material that is simply treated in a normal wastewater treatment plant following the process. Less energy is needed. Compared with conventional anodized rods, the energy saving is more than 50 percent.

All the materials used are harmless for the environment and health. This also makes it suitable for allergy sufferers. Eco-friendliness is combined with a high level of functionality and quality. Blue One is the world's first bluesign® tent. This strict environmental standard monitors the entire textile value-added chain. This makes Blue One a further step in the company's plans to be Europe's most ecological provider of outdoor equipment – a goal that Vaude wants to achieve by 2015.

Serniga 800
Alpin Schlafsack

Alpine sleeping bag

Jury
Judges panel

Sport
Silber

Sport
Silver

Er besitzt verschiedene Merkmale, die herkömmliche Schlafsäcke für hochalpine Touren oder Expeditionen nicht haben. Das Material ist recycelbar und trotzdem wasserdicht. Man kann den Schlafsack in sich selbst verpacken und braucht keinen zusätzlichen Packsack. Für kleinere Menschen kann man ihn verkürzen und damit die Füße noch wärmer halten. Richtig gut ist auch der Kragen im Halsbereich, der den Wärmeverlust reduziert. Warm, komfortabel und ökologisch.

It has various features that conventional sleeping bags for high-altitude tours or expeditions do not have. Materials have been used that are recyclable, but nonetheless water-resistant. The sleeping bag is packed by turning it in on itself, eliminating the need for any extra storage sack. For smaller people, it can be shortened, thus keeping their feet warmer. The collar around the neck area is a really good feature, reducing the loss of warmth. Warm, comfortable and ecological.

Hersteller
Manufacturer
Vaude Sport GmbH & Co. KG
D-88069 Tettnang

Design
Designer
Werksdesign
In-house design
Katrin Bojarski, Philipp Ziegler

Vertrieb
Distributor
Vaude Sport GmbH & Co. KG
D-88069 Tettnang

Serniga 800 ist ein High-performance-Schlafsack für Extremeinsätze wie Expeditionen und Hochtouren in sehr kalten Regionen. Er ist leicht, wasserabweisend und gleichbleibend warm durch die Trapezkammerkonstruktion, in der die Daune gleichmäßig verteilt bleibt.

Der Mumienschlafsack wurde nach dem Motto »reduce« und »reuse« entwickelt. Reuse: Er besitzt keinen extra Packsack. Das bedeutet weniger Materialverbrauch. Reduce: Der Fußsack dient zugleich als Packsack.

Nicht nur das innovative Packkonzept zeichnet den Serniga 800 aus, sondern auch eine neuartige Kapuzenöffnung. Diese lässt sich so eng zuziehen, dass nur ein kleines Atemloch übrig bleibt. Möglich ist dies durch den Trapezschnitt. Zudem verbessert ein dreidimensionaler Extrem-Wärmekragen die Isolation deutlich. Durch den geschwungenen Schnitt passt er sich optimal an die Konturen des Körpers an und verringert den Luftaustausch mit der Umgebung spürbar.

Für Menschen, die kleiner als 1,95 sind, kann das Fußende nach innen gestülpt werden, oder es lässt sich zur Aufbewahrung für Foto oder Camcorder nutzen. Wasserdichte Materialien sorgen dafür, dass Kopf- und Fußbereich nicht nass werden, wenn sie am Zelt anstoßen.

Der bluesign®-zertifizierte Schlafsack verbindet hohe Funktionalität und Komfort mit Ökologie. Statt in Asien werden die Daunen nun in Deutschland direkt vor Auslieferung befüllt. Das verkürzt nicht nur die Transportwege und verringert den CO_2 Ausstoß, sondern erhöht auch den Loft.

Serniga 800 is a high-performance sleeping bag for the sort of extreme conditions encountered on expeditions and mountaineering tours in very cold regions. It is light and water-resistant, and its trapezoid baffle construction, which evenly distributes the down filling, means that it is constantly warm.

The mummy sleeping back was designed according to the principles »reduce« and »reuse«. The reuse principle means it does not have any extra storage sack. This means that less material is used. The reduce principle means that the foot sack doubles up as a storage sack.

This innovative storage idea is not the only surprising feature of the Serniga 800. It also has a novel hood opening, which can be drawn so tightly closed that only a small breathing hole remains. The trapezoid design makes this possible. In addition, insulation is significantly improved by a three-dimensional extreme warmth collar. Its curved shape means that it adapts perfectly to the contours of the body, appreciably reducing the exchange of air with the environment.

For users who are shorter than 1.95 meters, the foot end can be turned inward, or it can be used for storing a camera or camcorder. Watertight materials ensure that the head and foot zones do not get wet if they come into contact with the wall of the tent.

This bluesign®-certified sleeping bag combines functionality and comfort with ecology. Instead of being filled in Asia, the sleeping bag is now filled with down in Germany, directly before being dispatched. This not only reduces transport distances and CO_2 emissions, but also increases the sleeping bag's loft or fluffiness.

VAUDE

HTX Lector 5800
Montainbike

Montain bike

Jury
Judges panel

Sport
Silber

Sport
Silver

Ästhetisch und formal schön. Dieses Fahrrad macht das technologische Wettrennen im überrüsteten Fahrrad-, Downhill- und Mountainbike-Markt optisch nicht mit, sondern versucht, über eine ganz saubere ästhetische Gestaltung zu reduzieren und nicht den Maschinenbaucharakter der Mechanikkomponenten hervorzuheben. Die Linienführung finden wir sehr schön durch die Schwarz-Weiß-Kombination, die auch in den Reifen fortgeführt wird. Flüssiger Rahmen. Gutes Preis-Leistungs-Verhältnis.

Tasteful and formally elegant. Optically, the bike remains aloof from the technological race in the over-equipped bicycle, downhill and mountain bike markets. Instead, it uses a well-executed and aesthetically pleasing design to play this down somewhat, and get away from the mechanical engineering character of the mechanical components. We find the lines well done, with the combination of black and white that is continued into the tyres. Flowing frame. Good value for money.

Hersteller
Manufacturer
Ghost Bikes GmbH
D-95652 Waldsassen

Design
Designer
Werksdesign
In-house design

Vertrieb
Distributor
Ghost Bikes GmbH
D-95652 Waldsassen

Das HTX Lector ist von allen Ghost Bikes das erfolgreichste. Mit keinem anderen wurden mehr Welt- und Europameisterschaften gewonnen.

Für 2011 wurde es noch besser und schneller gemacht. Ein neuer, noch leichterer Rahmen, gespickt mit allen technischen Raffinessen wie Postmount Direct – dadurch konnten beim Anbringen des Bremssattels zwei Schrauben eingespart werden – und E-Type Direct, wodurch der Umwerfer statt mit einer Schelle um das Sattelrohr direkt an der Hauptlagerung des Hinterbaus angebracht werden konnte. Vorteile sind neben Gewichtsreduktion vor allem die Verkleinerung des Hebels, die den Schaltvorgang schneller und direkter macht.

Für höheren Fahrkomfort, mehr Stabilität und Steifigkeit sorgen die bereits bewährten Technologien SCL Specific Chainstay Length – wobei die Länge der Kettenstrebe mit der Rahmengröße wächst – und Pressfit – hier werden die Lagerschalen direkt in das Tretlagergehäuse des Carbonrahmens eingepresst. Und nicht zu vergessen Tapered Headtube – dank dem nach unten verbreiterten Steuerrohr kann das Unterrohr breiter gebaut werden, was neben der Steifigkeit dem Lenkverhalten zugute kommt.

Der superleichte Carbonrahmen bietet trotz enormer Steifigkeit ein Höchstmaß an Komfort. Sattelstütze und Lenker sind ebenfalls aus dem leichten Kohlenstoff. So wiegt das Mountainbike lediglich 10,7 Kilo. Das Bike, das auch durch sein ästhetisches Design besticht, ist ideal für den sportlichen Marathon- und Cross-Country-Fahrer.

The HTX Lector is the most successful of all the ghost bikes. No other bike has won more world and European championships.

For 2011, it has been made even better and faster. A new, even lighter frame featuring every technical sophistication such as Postmount Direct – which saves two screws when attaching the brake calliper –and E-Type Direct, which allows the derailleur to be mounted directly on the frame, instead of mounting it around the seat tube using a shell. Apart from reducing weight, the main advantage of this is that leverage is reduced, making gear shifting faster and more direct.

Other tried and tested technologies provide for more riding comfort, stability and stiffness: SCL specific chainstay length, where chainstay lengths actually grow with their respective frame sizes, and Pressfit, where the bearing shells are pressed directly into the bottom bracket shell of the carbon frame. And then there is the tapered headtube, which broadens towards the bottom, allowing the down tube to be made wider. Apart from increasing stiffness, this improves the bike's steering.

Despite its extreme stiffness, the superlight carbon frame is extremely comfortable. The seat post and handlebars are also made of this light carbon. As a result, the mountain bike weighs just 10.7 kilos. With its outstandingly tasteful design, the HTX Lector5800 is the ideal bike for the sporty marathon and cross-country rider.

SCHWALBE
ROCKET RON
GHOST
HTX LECTOR

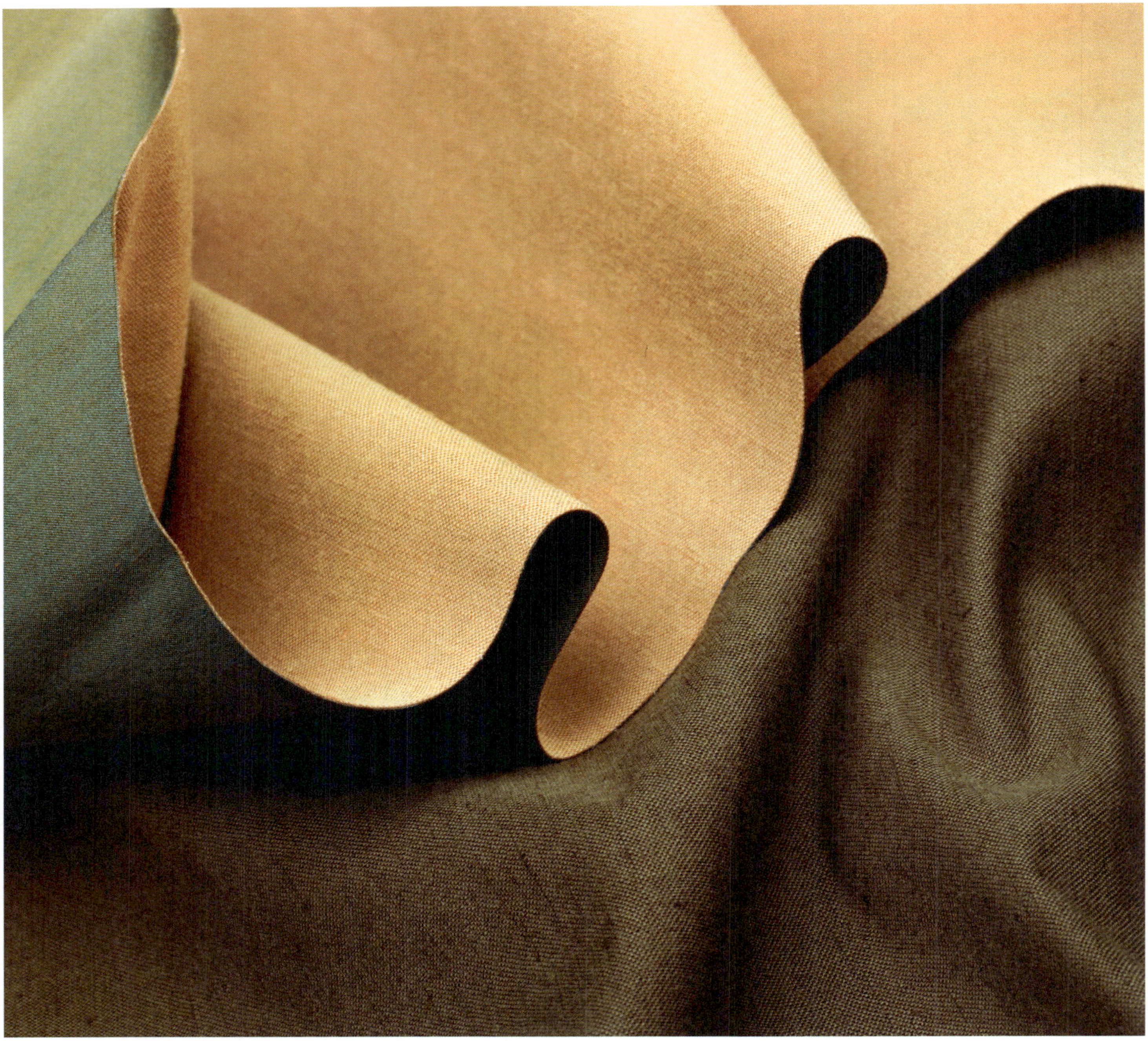

Energear™ Textilausrüstungstechnologie

Textile equipment technology

Jury
Judges panel

Sport
Silber

Sport
Silver

Außergewöhnliches Material. Wir finden die Funktionalität sehr überzeugend, weil wir glauben, dass es bestimmte Bereiche am Körper gibt, wo Wärme zur Unterstützung der Muskulatur zurückgeführt werden sollte. Die reflektierte Energie steigert somit die Leistung und das Wohlbefinden. Wir haben so etwas bisher noch nicht gesehen.

Extraordinary material. We were very impressed by its functionality. We feel that that are certain parts of the body to which warmth needs to be returned in order to support the muscular system. In addition, the reflected energy improves performance and enhances well-being. We've never seen anything like it.

Hersteller
Manufacturer
Schoeller Textil AG
CH-9475 Sevelen

Design
Designer
Werksdesign
In-house design

Vertrieb
Distributor
Schoeller Textil AG
CH-9475 Sevelen

Die völlig neue Textiltechnologie sorgt dafür, dass vom Körper abstrahlende Energie zurückgeführt wird. Eine speziell entwickelte Titan-Mineralien-Matrix bewirkt, dass vom Körper abstrahlende Wärmeenergie in Form von Ferninfrarotstrahlen vom Textil zurückprojiziert wird. Hierfür wurde die seit Jahrhunderten bekannte natürliche Eigenschaft bestimmter Mineralien, Wärmeenergiestrahlen zu reflektieren, genutzt. Ferninfrarotstrahlen sind Bestandteil der Infrarotstrahlung. Sie werden von der Sonne, bestimmten Steinen und Metallen oder auch vom menschlichen Körper abgestrahlt. Ihre therapeutische Wirkung wird in der asiatischen Heilmedizin bereits seit langem genutzt, da sie positiven Einfluss auf energetische Vorgänge im menschlichen Körper haben.

Durch die Reflexion der körpereigenen Infrarotstrahlung wird dem Körper zusätzliche Energie zur Verfügung gestellt. Der Wärmehaushalt wird optimiert, die Blutzirkulation gefördert und der Sauerstoffanteil im Blut erhöht, was sich positiv auf den Körper auswirkt: schnellere Aufwärmphase und mehr Leistung, weniger frühzeitige Ermüdung und bessere Regeneration, höhere Konzentrationsfähigkeit und gesteigertes Wohlbefinden.

Andere Gewebeeigenschaften wie etwa Atmungsaktivität, Wetterschutz oder Elastizität bleiben selbstverständlich erhalten.

Dieses innovative Gewebe lässt sich da einsetzen, wo körperliche Leistungsfähigkeit und Ausdauer gefragt sind. In der Bekleidung für den Bike- und Bergsport, zum Trekken und Wandern, beim Motorradfahren, in Arbeits- und Militärbekleidung, aber auch in City- und Streetwear.

This completely new textile technology ensures that the energy radiated by the body is recovered. The specially developed titanium-mineral matrix used in the textile reflects a body's far-infrared rays (FIRs) back to the wearer. This development is based on centuries-old knowledge about the natural ability of certain minerals to reflect far infrared rays. These FIRs are radiated by the sun, by certain stones and metals, and also by the human body. Their therapeutic effect has long been used in Asian healing methods, since they have a positive influence on energetic processes in the human body.

Reflecting the infrared rays radiated by the body makes additional energy available. The heat balance is optimized, blood circulation is stimulated, and the amount of oxygen in the blood increased, which has a positive effect on the body: warm-up phases are shortened, performance is improved, the body tires less easily and regenerates better, concentration is improved and well-being increased.

Other textile properties, such as breathability, weather protection and stretch, are of course unaffected.

This innovative textile can be used wherever physical fitness and stamina are needed. It can be used in apparel for cycling and alpine sports, for trekking and hiking and for motorcycling, as well as in work clothes and military gear, or indeed in city and streetwear.

Miau
Babyspielset

Set of toys for babies

Jury
Judges panel

Kids, Familie
Silber

Children, family
Silver

Wir empfinden alle drei Produkte als sehr stimmig, auch das Konzept der gemeinsamen Verpackung. Die Gestaltung entspricht dem Unternehmensprofil von Naef, sehr ästhetisch, nicht überfrachtet mit Farben. Es ist minimalistisch und besteht aus Naturmaterialien. Gerade in dieser Einfachheit finden wir das hervorragend. Eine sehr gute Idee von der Interaktion her sind die Augen, denn sie ziehen Kinder sofort in ihren Bann.

We feel that all three products make sense. It's also good that they are sold in one pack. Their design is in line with the Naef company profile – very tasteful, with the colours not overdone. It is minimalistic, made of natural materials. It is especially this simplicity that we think is excellent. In terms of interaction, the eyes are a very good idea. They are something that attract children's attention immediately.

Hersteller
Manufacturer
Naef Spiele AG
CH-4800 Zofingen

Design
Designer
Heiko Hillig
CH-4314 Zeiningen

Rund 15.000 Stunden spielen Kinder in ihren ersten sechs Lebensjahren. Deshalb ist es nicht gleichgültig, womit sie spielen.

Gerade für die Entwicklung des Babys ist Fühlen, Hören, Sehen und Greifen von zentraler Bedeutung. Miau bietet mit seiner babygerechten Form, seinen unbedenklichen Materialien und seinem Gewicht dem Baby einen idealen Start in die Welt des Spiels.

Das dreiteilige Babyspielset besteht aus einem Greifling, einer Kinderwagen- und einer Schnullerkette. Es besticht durch seine reduzierte und hochwertige Gestaltung ebenso wie durch das humorvolle Design des stilisierten Katzengesichts, dessen grünen Augen sich kein Baby entziehen kann. Die Spielzeuge bestehen aus naturbelassenem Ahornholz, wurden auf Festigkeit geprüft und mit speichelfesten und giftfreien Farben hergestellt.

Der Greifling schult nicht nur die kindlichen Greif- und Fingerfertigkeiten, sondern das Baby kann ihn auch, wenn es Zähne bekommt, bedenkenlos mit dem Mund erkunden. Dann wird der Greifling zum Kauring und erleichtert das Zahnen.

Am Kinderwagen ziehen die grünen Katzenaugen der Kette die Babyblicke auf sich, und die kleinen Hände bringen das Katzengesicht zum Rotieren. Dank der Schnullerkette geht der Schnuller nicht verloren, sondern bleibt immer in Reichweite.

In the first six years of their lives, children spend roughly 15,000 hours playing. So the question of what they play with is not a minor one.

Especially for babies' development, feeling, hearing, seeing and grasping are crucially important. Made of harmless materials, and with a shape and weight that is just right for babies, Miau gives baby an ideal start to the world of play.

This three-part set of baby toys comprises a rattle, a chain for the pram, and a clip and chain for a dummy. Its reduced yet high-quality design is impressive, as well as the humorous motif and a stylized cat's face, whose green eyes no baby will be able to resist. The toys are made of untreated maple, are tested for strength and produced using saliva-resistant, non-toxic dyes.

The rattle not only trains the child's grasping and manipulative skills. When the baby starts teething, it can also put the rattle in its mouth without any fear of harm. In this way, the rattle becomes a teething ring.

The green cat's eyes on the chain attached to the pram attract the baby's attention, and the baby's small hands can make the cat's face rotate. Thanks to the chain and clip, a dummy will not get lost, but will remain constantly within reach.

180
181

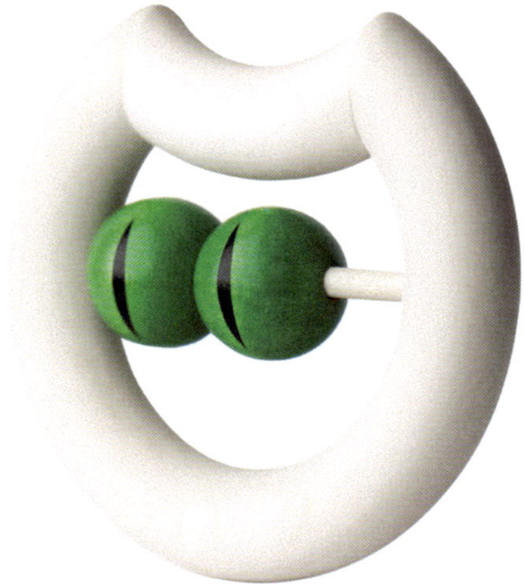

Jury
Judges panel

3plus
Farbstift

Crayon

Kids, Familie
Silber

Children, family
Silver

Schönes Produkt für kleine Kinder. Die Stifte liegen gut in der Hand, Aufbereitung, Verpackung und Darstellung sind sehr gelungen. Einfach und gut. Sehr sinnvoll: ein Anspitzer ist auch im Set enthalten.

Lovely product for small children. The crayon fits snugly in the hand, and it is very attractively finished, packed and presented. Simple and good. It also makes a lot of sense that a sharpener comes with it.

Hersteller
Manufacturer
C. Josef Lamy GmbH
D-69123 Heidelberg

Design
Designer
IPDD GmbH & Co. KG
D-70173 Stuttgart

Vertrieb
Distributor
C. Josef Lamy GmbH
D-69123 Heidelberg

Die Fähigkeit, ein Schreibgerät ergonomisch richtig greifen und führen zu können, wird bereits weit vor der Einschulung entwickelt. Beim Malen und Zeichnen üben Kinder spielerisch und mit viel Spaß die motorischen Fähigkeiten ihrer Hände und Finger – eine der wichtigsten Grundlagen, um später in der Schule schreiben zu lernen.

Deshalb ist es von großer Bedeutung, für jede Entwicklungsstufe den richtigen Farbstift zur Verfügung zu stellen. Im Alter von drei Jahren greifen Kinder einen Stift noch sehr grob mit der vollen Hand. Sie führen ihn oft mit viel Kraft über den Untergrund. Genau für diese Anforderungen ist der 3plus entwickelt und gestaltet.

Der Farbstift für Kinder ab drei Jahren ist Farb-, Aquarell- und Wachsmalstift in einem. Durch seine ergonomische quadratische Form mit den abgerundeten Kanten und seine kompakte Länge können ihn die kleinen Künstler besonders gut auch mit der ganzen Hand greifen.

Der 3plus zeichnet sich durch hohe Farbbrillanz, außergewöhnliche Deckkraft und einen weichen Abstrich aus. Die Kinder können mit ihm auf allen glatten Oberflächen malen und ihn auch mit Wasser vermischen.

Der Farbstift besteht aus hochwertigem Zedernholz, das aus nachhaltig bewirtschafteten Wäldern (PEFC-Siegel) stammt. Er ist besonders bruchfest, seine extradicke – 10 Millimeter – Mine hält ebenfalls größten Belastungen stand. Und die matte Lackierung sorgt für einen sicheren Griff.

The ability to grip and guide a writing implement in the ergonomically correct way is something that is developed well before starting school. When colouring and drawing, children practice the motor skills of their hands and fingers, in a playful, fun way. This is one of the most important foundations for learning to write later at school.

This is why it is extremely important that the right crayons are available for each stage of development. At the age of three, children still grasp a crayon very roughly, using their entire hand. They are often very heavy-handed when guiding the crayon. It is precisely for these requirements that 3plus has been developed and designed.

This crayon for children aged three and above is a combination of crayon, aquarelle pencil, and wax crayon. With its ergonomic square section with rounded corners and its compact length, it is a crayon that tiny artists can grasp especially easily, also with their entire hand.

3plus offers unusually brilliant colours and opacity, as well as a soft lead. Children can use it to draw on smooth surfaces of all kinds, and can also use it in combination with water.

The crayon is made of high-quality cedar grown in sustainably managed forests (PEFC seal). It is especially sturdy, and its extra thick lead (10 millimetres) can also withstand very tough treatment. The crayon's matt coating ensures a good grip.

LAMY

Jury
Judges panel

Aquaplus Deckfarbkasten

Box of opaque watercolour paints

Kids, Familie
Silber

Children, family
Silver

Sehr leicht verständlich wird der Farbkreis von Johannes Itten erklärt. Und das in Lamy-typischer Art: typografisch gut aufbereitet als kleines Heftchen. Auf dem Deckel und der Verpackung ist der Farbkreis ebenfalls abgedruckt. Ein anderes schönes Detail ist das Namensfeld. Die Teile, die sich bewegen, sind ganz eindeutig symbolisiert. Man weiß genau, wo man drücken muss. So ist selbst bis hin zum Auswechseln der Farbe alles klar, eindeutig und selbsterklärend.

Johannes Itten's colour sphere is explained in a very easy way. The way this is done is typical of Lamy: in a typographically well designed booklet. The colour sphere is also displayed on the lid and the packaging. The space for the owner's name is another nice detail. The parts that move are clearly indicated. You know exactly where you have to press. Right down to replacing colours, everything is crystal clear and self-explanatory.

Hersteller
Manufacturer
C. Josef Lamy GmbH
D-69123 Heidelberg

Design
Designer
Franco Clivio
CH-8703 Erlenbach

Vertrieb
Distributor
C. Josef Lamy GmbH
D-69123 Heidelberg

Der Aquaplus mit Farbkreis nach Johannes Itten macht Farbenlehre spielerisch erlebbar. Hier können Kinder ausprobieren und entdecken, wie sich Farben mischen lassen. Deshalb kann er ideal im Schulunterricht eingesetzt, aber natürlich auch nur zum Malen verwendet werden.

Der Farbkasten wurde ganz im Sinne der Farbenlehre von Johannes Itten, künstlerischer Leiter am Bauhaus in Weimar, entwickelt. Dabei handelt es sich um ein Farbenmodell, das das Mischprinzip der Farben unterschiedlicher Ordnungen darstellt. Den Ausgangspunkt bilden die Farben erster Ordnung – Rot, Gelb und Blau. Aus ihnen lassen sich alle zwölf Farben des Farbkreises mischen.

Mit Hilfe des beiliegenden Leporellos lernen Kinder und Jugendliche den Zusammenhang und die Entstehung von Farben kennen und können jederzeit selbst durch Mischung verschiedener Farben den von ihnen gewünschten Farbton erzeugen. Als Orientierungshilfe oder um das Mischergebnis direkt überprüfen zu können, ist auf dem Deckel noch einmal der Itten-Farbkreis abgebildet.

Die Farbtöne des Farbkastens sind rein und brillant und zeichnen sich durch eine hochergiebige konstante Farbqualität aus. Aquaplus besitzt im Deckel dreizehn leicht zu reinigende Mischfelder inclusive einem Lamy-Sternmischfeld. Die einzelnen Farbschalen sind auswechselbar – bis zu 24 Farben können einzeln nachgekauft werden. Der Deckel mit einem stabilen Doppel-X-Scharnier lässt sich abnehmen.

Aquaplus, with a colour sphere influenced by Johannes Itten, offer a playful introduction to the theory of colour. Children can experiment and discover how different colours can be mixed. This makes it ideal for the classroom, though it can of course simply be used for painting in leisure time.

The paint set was developed according to the theory of colour put forward by Johannes Itten, who taught at the Bauhaus in Weimar. This theory is a model that presents the principle of mixing colours belonging to different orders. It starts from the colours of the first order – red, yellow and blue. They can be used as the basis for mixing the twelve colours of the colour sphere.

Using the foldout brochure that comes with the set, children and young people can get to know the relationship between different colours and how colours are created. They can mix their own colours to produce the tone they are looking for. As a guide, and as a way of directly checking the mixing result, Itten's colour sphere is displayed once more on the lid.

The colours in the set are pure and brilliant, and yield a long-lasting, constantly good quality of colour. The lid of aquaplus is furnished with thirteen easy-to-clean mixing trays., including a Lamy mixing tray. The individual pans can be replaced, and up to 24 colours can be purchased individually. A sturdy double-x hinge allows the lid to be removed.

LAMY
DECKWEISS • CHINESE WHITE
Made in Germany
www.lamy.com
uaplus
24

Hersteller
Manufacturer
Bosch Thermotechnik GmbH
D-73249 Wernau

Design
Designer
Werksdesign
In-house design
Fusion-x Community Inhouse
und
Intuity Media Lab GmbH
D-70176 Stuttgart

Vertrieb
Distributor
Bosch Thermotechnik GmbH
Junkers Deutschland
D-73243 Wernau

JUNKERS HOME BOSCH

Junkers Home
Heizungsfernsteuerung per App

Heating control app

Architektur
Gold

Architecture
Gold

186
187

Junkers Home
Heizungsfernsteuerung per App

Heating control app

Architektur
Gold

Architecture
Gold

Jury
Judges panel

Eine hervorragende Anwendung. Diese App begeistert uns. Sie ist selbsterklärend und einfach zu bedienen. Sehr gut gestaltet – die Reduktion auf das Notwendige. In den jeweiligen Räumen interessiert nur die Temperatur und genau die kann man hier einstellen. Die Bedienung der Junkers-App ist intuitiv auch deshalb, weil sie der intuitiven Bedienung der anderen Apps folgt. Man braucht keine Bedienungsanleitung.

An excellent application. We love it. It is self-explanatory and easy to operate. Very well designed, especially the way it has been reduced to what is necessary. If I have central heating, all I want to know is how warm it is in my flat, and this is precisely what this app allows me to set. The Junkers app is so intuitive to use because it follows the same intuitive principles as other apps. I don't need any instruction booklet.

Hersteller
Manufacturer
Bosch Thermotechnik GmbH
D-73249 Wernau

Design
Designer
Werksdesign
In-house design
Fusion-x Community Inhouse
und
Intuity Media Lab GmbH
D-70176 Stuttgart

Vertrieb
Distributor
Bosch Thermotechnik GmbH
Junkers Deutschland
D-73243 Wernau

Früher Feierabend machen und die Heizung bereits auf dem Heimweg aktivieren oder seinen Urlaub verlängern und das direkt von der Skipiste seiner Heizung mitteilen können. Eine Zukunftsvision?

Möglich ist das mit der App Junkers Home. Damit können von überall auf der Welt via Internet die Schaltzeiten neu eingestellt oder die Betriebsart gewechselt werden – intuitiv und ohne Bedienungsanleitung, wie der Benutzer es von seinem iPhone gewohnt ist.

So ist die Wohnung beim Betreten immer schön warm. Und wenn niemand zuhause ist, spart die Heizung Energie und damit Geld. Bequemer kann Heizungsregelung kaum sein.

Zudem zeigt Junkers Home Wartungs- oder Störungsmeldungen an. So kann der Heizungsfachmann gegebenenfalls auch von unterwegs informiert werden. Ist die Heizung mit einer Solaranlage kombiniert, kann sich der Besitzer mit der App außerdem den Solarertrag als Diagramm anzeigen lassen.

Die App kann im iTune App Store von Apple heruntergeladen werden. Kompatibel ist die mobile Anwendung derzeit mit iPhone, iPad und iPod touch.

Damit die Heizung die Befehle versteht, braucht es noch einen Vermittler: das Internet-Kommunikationsmodul MB 100-LAN von Junkers, ein sogenanntes Gateway. Die kleine weiße Box wird vom Heizungsfachmann an die Junkers-Heizungsanlage angeschlossen und mit dem Internet-Router verbunden. Und fertig ist die Fernsteuerung für mehr Komfort und Kostenersparnis.

Leaving work early and activating the central heating on the way home, or prolonging your holiday and letting your heating know directly from the piste. Pie in the sky?

It is now reality with the Junkers Home app. Now you can set your central heating from anywhere in the world via the internet, and you can do so intuitively and without any instructions, just as iPhone users would expect of an app.

Now you can be sure of a warm home whenever you return. And if nobody is at home, the central heating saves energy and money. Controlling the central heating has scarcely been easier.

Junkers Home also lets the user know when a service is due or when there is a malfunction. In this way, it is also possible for the heating technician to be informed via his smartphone. If the central heating is combined with a solar thermal system, the app can present the owner with a diagram showing how much yield he has had from the sun.

The app can be downloaded from the Apple iTune App Store. The mobile application is currently compatible with an iPhone, iPad and iPod touch.

For the central heating to be able to understand the commands, a go-between is needed. This is provided in the form of the Junkers MB 100-LAN internet communication module, which acts as a gateway. The heating technician connects this small white box to the Junkers heating system and to an internet router. And that is all that is needed for a remote control that provides added convenience and cost savings.

PM
15
18
21
auto
Info
Setting

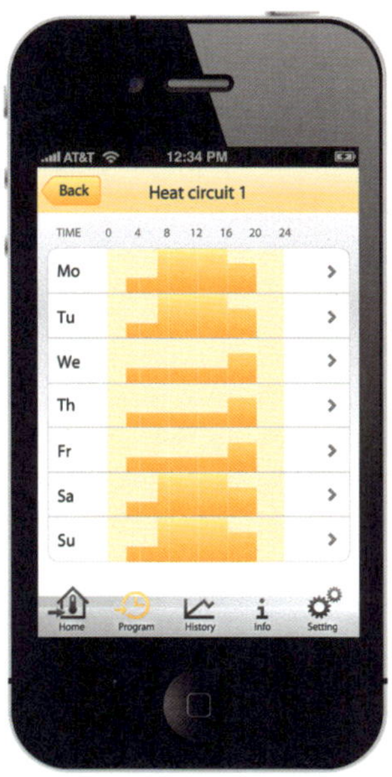
AT&T
12:34 PM
Back
Heat circuit 1
TIME 0 4 8 12 16 20 24
Mo
Tu
We
Th
Fr
Sa
Su
Home
Program
History
Info
Setting

AT&T
12:34 PM
Back
Temperature Levels
30°C
25°C
18°C
13°C
6°C
13°C
Home
Program
History
Info
Setting

Vitotrol 300 RF
Funkfernbedienung

Wireless remote control

Jury
Judges panel

Architektur
Silber

Architecture
Silver

Schöner Fokus auf die wesentlichen Elemente, dadurch ist die Funkfernbedienung einfach zu bedienen. Weitergehende Funktionen befinden sich nicht störend im Vordergrund, sondern in Untermenüs. Das Display ist kontrastreich, weiß auf schwarz mit wenigen Farben, nur dort, wo sie Sinn ergeben. Perfekter Bedienkomfort, sehr schön im Wohnbereich integrierbar, hervorragende Lösung für ein technisches Produkt.

Nice focus on the essentials, which makes the remote control easy to operate. Other functions do not get in the way, but can be found in sub-menus. High-contrast display, white on black with very few colours. Colour is only used where it makes sense. Perfect operating convenience, harmonizes excellently with living areas. Excellent solution for a technical product.

Hersteller
Manufacturer
Viessmann Werke GmbH & Co. KG
D-35108 Allendorf

Design
Designer
Phoenix Design GmbH & Co. KG
D-70376 Stuttgart

Vertrieb
Distributor
Viessmann Werke GmbH & Co. KG
D-35108 Allendorf

Mit Vitotrol 300 RF können alle Funktionen der Heizung bequem vom Sofa aus bedient und kontrolliert werden. Möglich wird diese sehr komfortable Benutzung unabhängig von lästigen Kabeln durch den Akkubetrieb und die Funkkommunikation mit der Heizungssteuerung.

Über ein großzügiges Touch-Display mit intuitiv verständlicher grafischer Benutzeroberfläche ist die neu entwickelte Wohnraum-Funkfernbedienung ganz einfach zu bedienen. Sie verbindet leicht nachvollziehbare logische Strukturen mit bereits gelernten Bedienmustern, die sich an den Bedürfnissen der Benutzer orientieren. Das Handbuch muss dabei nicht zu Rate gezogen werden.

Auf dem Home-Bildschirm sind klar und übersichtlich die Funktionen für die unmittelbarsten Bedienvorgänge angeordnet. Weitergehende Funktionen befinden sich in eindeutig und logisch strukturierten Untermenüs.

Neben ihrem perfekten Bedienkomfort besticht Vitotrol 300 RF durch die Konzentration auf das Wesentliche und ihre hochwertige Anmutung. Durch ihre klare und moderne Gestaltung integriert sich die Funkfernbedienung hervorragend in das Wohnumfeld.

Das perfekte Hochglanzfinish des Displaybereichs setzt einen kontrastreichen Akzent in dem hochwertigen silbernen Gehäuse. Das handschmeichlerische Gehäuse kann wahlweise nicht nur in einem Tischständer aufgestellt, sondern auch in einer Halterung an die Wand montiert werden.

Vitotrol 300 RF can be used to operate and control all heating-system functions from the comfort of your sofa. The fact that this can be done so conveniently, without bothersome cables, is due to rechargeable batteries and wireless communication with the heating controls.

A generously dimensioned touch display, with a graphic user interface that is intuitive to understand, makes this newly developed wireless remote control for the living room very easy to operate. It combines easy-to-understand logical structures with already learned operating patterns, which orient to the needs of the user. There is no need to consult the operating manual.

The functions for the most basic operating processes are arranged clearly and logically on the home screen. Other functions can be found in sub-menus, which are clearly and logically structured.

Apart from perfect operating convenience, what stands out about Vitotrol 300 RF is its focus on the essentials and its high-quality appearance. Its clear and modern design means that the remote control harmonizes excellently with the living environment.

The perfect high-gloss finish of the display contrasts with the stylish silver housing. A joy to touch, the housing can be placed on a table stand or mounted on a wall bracket.

13:42
14°
20.5°
Menü
VIESSMANN

So 11.04.10
20:30
8,5°C
20,5°C
Menü
Räume
Licht
Jalousie
Kamera
Türen und Fenster
Türkom-munikation
Energie-verbrauch
Alarm
Diagramm
Wetter-prognose
Wetter-station
Szenen
Mail
News
Musik
Außen-bereich
Menü
Favoriten
MyTouch
Notiz
System
Visu
GIRA Control 9

Gira Control 9 Client
Anzeige- und Bedieneinheit

Display and control unit

Jury
Judges panel

Architektur
Silber

Architecture
Silver

Die Anzeige- und Bedieneinheit ist ganz in der Tradition des Unternehmens: sehr gut, sachlich, aber auch modern gestaltet mit einem besonderen Fokus auf das Interface-Display, das durch eine klare und leicht ablesbare Typografie sowie eine intuitive Menüführung besticht. Die Symbolik und die Ikonographie sind gut und ansprechend, nicht zu technisch, da das Gerät im Wohnbereich verwendet wird. Schöner Materialmix. Ein technisch sehr ausgeklügeltes Produkt.

The display and control unit is just as you would expect from this company: very good, sober yet modern in design, with a special focus on the interface display, which stands out for its clear, easy-to-read typography and intuitive navigation. The symbols and iconography are good and appealing, not too technical, since the device is used in living areas. Beautiful mix of materials. A technically very sophisticated product.

Hersteller
Manufacturer
Gira Giersiepen GmbH & Co. KG
D-42477 Radevormwald

Design
Designer
Werksdesign
In-house design

Interface-Design
Interface designer
Schmitz
Visuelle Kommunikation
Hans Günter Schmitz
D-42103 Wuppertal

Vertrieb
Distributor
Gira Giersiepen GmbH & Co. KG
D-42477 Radevormwald

Der Gira Control 9 Client lässt sich einfach und bequem mit einem Finger steuern. Mit dieser PC-basierten Kontroll- und Steuereinheit für den Gira HomeServer3 oder den FacilityServer kann die gesamte Gebäudetechnik basierend auf dem KNX/EIB System zentral überwacht und gesteuert werden. Außerdem lässt sich damit auch das Internet nutzen und so beispielsweise morgens in Erfahrung bringen, ob man einen Regenschirm mitnehmen muss.

Der brillante Touchscreen garantiert eine haptisch angenehme und einfache Bedienung. Das klar gestaltete Interface stellt die gesamte Gebäudetechnik übersichtlich und anschaulich dar. Sämtliche Funktionen sind innerhalb von zwei Ebenen erreichbar. Die verständliche und intuitive Menüführung sorgt für einen schnellen Zugriff auf die unterschiedlichsten Bereiche wie Licht, Jalousien, Lüftung, Heizung und Hausgeräte.

Darüber hinaus können zusätzliche Funktionen bedient werden, wie Lichtszenen speichern und abrufen, Musik hören, Internetanwendungen nutzen und E-Mails anzeigen, Alarmmeldungen darstellen sowie Diagramme abrufen, die Verbrauchsdaten erfassen und auswerten. Durch die Integration von Lautsprecher und Mikrofon lässt sich der Gira Control 9 Client auch für die Türkommunikation nutzen.

Das Gerät wird in einer Unterputzdose in die Wand montiert. Der Rahmen besteht aus Aluminium und Glas, das in Schwarz, Weiß, Mint oder Umbra erhältlich ist.

Über eine App kann die gesamte Gebäudetechnik auch mobil bedient werden – über iPhone, iPod touch oder iPad, via GSM, UMTS oder WLAN.

Gira Control 9 Client can be controlled simply and comfortably using just one finger. This PC-based regulation and control unit for the Gira HomeServer 3 or the FacilityServer can be used to monitor and control any building automation technology based on the KNX / EIB system. It can also be used to access the web, and in this way to find out in the morning whether you need to take an umbrella with you.

Pleasant to the touch, the brilliant touch screen guarantees simple operation. The clearly structured interface presents the entire building automation technology in a clear way. All functions can be accessed within two levels. The understandable, intuitive navigation ensures fast access to diverse areas such as lighting, blinds, ventilation, heating and household appliances.

Apart from that, there are additional functions that can be controlled such as lighting sequences, listening to music, using internet functions and displaying emails, showing alarm reports and calling up diagrams that record and evaluate consumption data. An integrated loudspeaker and microphone allow Gira Control 9 Client to be used as an intercom for door communication.

The device is flush mounted on the wall. Its framework is made of aluminium and glass, which is available in white, mint or umbra.

An app allows the entire building automation technology to be controlled while on the move, using an iPhone, iPod touch or iPad, via GSM, UMTS or WLAN.

Giro water
Kaminofen mit Wasserwärmetauscher

Wood-burning stove with hot-water heat exchanger

Jury
Judges panel

Architektur
Silber

Architecture
Silver

Eine klare, saubere, architektonische Grundstruktur eines Kaminofens, der durch den Wasserwärmetauscher sehr energieeffizient und ökologisch sinnvoll ist. Schnörkellos, zurückhaltend und funktional.

A clear, neat, architectural basic structure of a wood-burning stove, whose hot-water heat exchanger makes it very energy efficient and ecologically beneficial. No frills, unassuming and functional.

Hersteller
Manufacturer
Wodtke GmbH
D-72070 Tübingen

Design
Designer
Werksdesign
In-house design

Vertrieb
Distributor
Wodtke GmbH
D-72070 Tübingen

Giro water bietet eine elegante Symbiose aus Feuer und Wasser. Neben der angenehmen Wärmeabstrahlung mit dem faszinierenden Flammenspiel gibt der Kaminofen gleichzeitig circa 70 Prozent der gewonnenen Energie über seinen Wasserwärmetauscher an Pufferspeicher und Heizungsanlage ab: CO_2-neutrales Heizen mit Anbindung an das Zentralheizungssystem – die Basis für Energie-Effizienz und Nachhaltigkeit.

Der Nutzer erkennt über eine rot leuchtende LED im Lagerfach, wann der Pufferspeicher genügend aufgeheizt ist und er nicht mehr Holz nachzulegen braucht. So wird die eingesetzte Energie sinnvoll, sparsam und effizient genutzt. Durch seine hohe Wasserleistung ist der Kaminofen in Kombination mit Solar auch hervorragend für Niedrigenergie- oder Passivhäuser geeignet.

Eine weitere zukunftsweisende technologische Innovation der Kaminofenwelt ist die Raumluft-unabhängige Luftzuführung. Dadurch kann der Kaminofen auch speziell in modernen, hoch wärmegedämmten Häusern mit kontrollierter Wohnraumlüftung eingesetzt werden.

Die optimale Feuerraumgeometrie in Verbindung mit der Wodtke-Thermoregelung mit automatischer Verbrennungsluftführung sorgt für einen schadstoffarmen Abbrand. Mit der AWS-Scheibenspülung wird Sekundärluft als »Luftvorhang« entlang der Scheibe nach unten zur Flamme geführt. So bleibt die Sichtscheibe länger sauber.

Neben seiner Feuer-Ästhetik besticht Giro water auch durch seine klare Formensprache. Das bevorzugte Material Stahl verbindet sich perfekt mit intelligenter Funktionalität.

Giro water is an elegant symbiosis of fire and water. Apart from the pleasant radiated heat with the fascination of flickering flames, the stove passes on roughly 70 percent of the heat generated to the buffer storage unit and heating system, via a heat exchanger. CO_2-neutral heating that is connected to the central heating system – a basis for energy efficiency and sustainability.

Via a red LED in the storage compartment, the user can see when the buffer storage unit has been heated up sufficiently, and there is no need to add further wood. In this way, energy is used sensibly, economically and efficiently. When the stove is combined with solar collectors, its high water capacity makes it eminently suitable for low-energy or zero-energy houses.

One further pioneering innovation in wood-burning stove technology is the room air-independent air supply. Above all, this allows the stove to be used in modern, extremely insulated houses in which ventilation is controlled.

The optimum geometry of the combustion chamber, combined with Wodtke thermo control with its automatic supply of combustion air, ensures that combustion is low in pollutants. AWS glass wash guides secondary air downward over the glass pane to the fire, like a kind of »air curtain«. This keeps the glass pane clean for longer.

Giro water stands out not only for the pleasure of watching a burning fire, but also for its clear formal expression. It is made mainly of steel, which combines perfectly with intelligent functionality.

RICHARD NEUTRA

Design
Designer
Atelier Brückner GmbH
D-70376 Stuttgart

KULTUREN DER WELT ATELIER BRUECK-NER

Kulturen der Welt
Rautenstrauch-Joest Museum Köln
Museumsgestaltung

Museum design

Public Design
Gold

Public design
Gold

Kulturen der Welt
Rautenstrauch-Joest Museum Köln
Museumsgestaltung

Museum design

Public Design
Gold

Public design
Gold

Jury
Judges panel

Ausgesprochen schöne Darstellung von verschiedenen Kulturen. Sie ist interaktiv, die Leute können sich hineinfühlen, die Dinge anschauen, anfassen, mitspielen, erfahren, erfassen und begreifen, auf eine Art und Weise und in einer Formschönheit, wie wir das bisher noch nicht gesehen haben. Herausragend.

Exceptionally good presentation of different cultures. It's interactive, people can feel their way in, look at things, touch them, play with them, experience, grasp and understand them. They do so in a way, and in a formally elegant setting, that we have never seen before. Excellent.

Design
Designer
Atelier Brückner GmbH
D-70376 Stuttgart

Das innovative Ausstellungskonzept verzichtet auf die übliche Einteilung in geografische Großräume und greift stattdessen Themen auf, deren Fragestellungen je nach Kulturkreis unterschiedlich beantwortet werden. Es bietet einen szenografisch inszenierten Parcours, in den rund 2.000 Exponate eingebettet sind.

Die Besucher erleben die Ausstellung als dynamische Abfolge einzelner Kapitel, die ihnen einen intuitiven Zugang zu den Kulturen der Welt eröffnen.

Innerhalb des ersten Hauptkapitels »Die Welt erfassen« werden vier Unterthemen visualisiert: »Begegnung und Aneignung: Grenzüberschreitungen«, »Der verstellte Blick: Vorurteile«, »Die Welt in der Vitrine: Museum« und »Ansichtssachen?!: Kunst«.

Der zweite Ausstellungsbereich »Die Welt gestalten« vermittelt in fünf Unterabteilungen Einblicke in unterschiedliche Formen der Lebensgestaltung: »Lebensräume, Lebensformen: Wohnen«, »Der Körper als Bühne: Kleidung & Schmuck«, »Der inszenierte Abschied: Tod und Jenseits«, »Vielfalt des Glaubens: Religionen« und »ZwischenWelten: Rituale«.

In der Abfolge der abwechslungsreich inszenierten Themenräume, die mit unterschiedlichen Materialien, Lichtstimmungen und Raumkompositionen arbeiten, entfaltet sich ein authentisch erlebbares, ethnologisches Panoptikum. Die Besucher werden direkt angesprochen, fühlen sich eingebunden in die museale Erlebniswelt und sind aufgefordert zum Kulturvergleich: die eigene Kultur als eine mögliche Variante unter vielen gleichwertigen Möglichkeiten zu begreifen.

This innovative exhibition concept does without the usual partition into geographical regions. Instead, it takes up themes whose issues are dealt with differently from culture to culture. It takes visitors on a scenographic route in which some 2,000 exhibits have been integrated.

Visitors experience the exhibition as a dynamic sequence of individual chapters that give them an intuitive insight into the world's cultures.

The first main chapter, »Grasping the World,« visualizes four sub-themes: »Encounter and assimilation: border crossings«, »Distorted perception: prejudice«, »The world in a showcase: museum« and »Viewpoints?! Art«.

The second exhibition area, »Shaping the world«, comprises five sub-departments that provide insights into the different ways in which people shape their lives. »Living space, ways of living: the home«, »The body as a stage: clothing and jewellery«, »Staging a farewell: death and the hereafter«, »Diversity of belief: religion« and »In-between worlds; rituals«.

As the visitor moves from one theme space to the next, each of which works with different materials, light moods and room compositions, an authentic ethnological panoptikon unfolds. Visitors are directly involved, and feel drawn in to the museum's various worlds. They are also invited to compare cultures, to understand their own culture as one possible variant among many equally valid possibilities.

DER VERSTELLTE BLICK

CERN-Universe of Particles Genf
Ausstellungsgestaltung

Exhibition design

Jury
Judges panel

Public Design
Silber

Public design
Silver

Ein hochwissenschaftliches Thema, das in einer Ausstellungsgestaltung umgesetzt ist und über mehrere Sinne leicht begreifbar gemacht wird. Man begibt sich hinein und es wird auf kompaktem Raum multivisuell und multimedial sowie auch interaktiv das Thema illustriert. Hier wird ein sehr verschlossenes, hochwissenschaftliches Thema auf sehr moderne Art begreifbar gemacht.

The subject that has been translated into an exhibition design is a highly scientific one, made easily comprehensible via all the senses. You immerse yourself in the subject, and in a compact space it is illustrated in a multivisual, multimedia-based, interactive way. In this exhibition, a secretive, highly scientific topic is made understandable in a very modern way.

Design
Designer
Atelier Brückner GmbH
D-70376 Stuttgart

Im CERN – Conseil Européen pour la Recherche Nucléaire – , dem weltweit größten Forschungszentrum auf dem Gebiet der Teilchenphysik, erforschen die weltbesten Physiker die Grundlagen unserer Existenz.

Das CERN-Besucherzentum hat die Aufgabe, interessierten Laien einen emotionalen Zugang zu komplexer wissenschaftlicher Forschungsarbeit zu bieten. Der Besucher betritt kosmische Welten: Interaktive Ausstellungselemente, sphärisch gestaltete Displays und eine Main Show, die jede halbe Stunde den Raum multimedial bespielt, stellen die grundlegende Forschungsarbeit vor, die im CERN geleistet wird.

Das Designkonzept basiert auf einer runden, sphärischen Formensprache, die die bestehende Globe-Architektur als Metapher eines unbegrenzten Raumes interpretiert. Mikro- und Makrokosmos, die beiden diametralen Forschungsbereiche, die in CERNs Fokus liegen, vereinen sich im Raumbild. Der Besucher taucht ein in eine Welt ohne Maßstäbe, er befindet sich zugleich auf subatomarer Ebene und in den unermesslichen Weiten des Sonnensystems.

Geschickte Material- und Lichtwahl sowie eine dreidimensional konzipierte Soundlandschaft führen zu einem nachhaltigen Raumerlebnis. Nichts stellt sich dem herumschweifenden Blick entgegen. Die Technik ist integrierter Bestandteil der sphärischen Formen. Computer-Interfaces, Beamer und Lautsprecher sind eingebettet und werden nicht per se wahrgenommen.

At CERN – Conseil Européen pour la Recherche Nucléaire – the world's largest research centre for particle physics, the world's best physicists research into the building blocks of our existence.

The CERN visitor centre's task is to offer the interested general public emotional access to complex scientific research work. Visitors enter cosmic worlds. Interactive exhibition elements, spherically formed displays and a main multimedia show that is shown every half an hour present the basic research work that is done at CERN.

The design concept is based on a round, spherical formal expression, which interprets the existing globe architecture as a metaphor of infinite space. The microcosmos and macrocosmos, the two diametrically opposed areas of research that are the focus of work at CERN, come together in the image space. Visitors immerse themselves in a world without benchmarks, moving both on a subatomic level and in the vast reaches of the solar system.

A clever choice of materials and lighting, as well as a soundscape designed to be three-dimensional, result in a lasting experience of space. Nothing stands in the way of the visitor's gaze as it moves around the room. The equipment is an integral part of the spherical shapes. Computer interfaces, projectors and loudspeakers are embedded, and are thus inconspicuous.

Boxer
Inflatables (aufblasbare Raumelemente)

Inflatables (inflatable room structures)

Jury
Judges panel

Public Design
Silber

Public design
Silver

Das System als Ganzes bietet eine große Vielfalt an verschiedenen Möglichkeiten. Es ist organisch, schön und multifunktional. Für Events ist es sehr modern, aber leicht und lässt sich branden. Es beginnt mit einem einfachen Grundmodell und hat endlose Möglichkeiten. Der Kreativität sind keine Grenzen gesetzt.

The system as a whole offers a great variety of different possibilities. It is organic, elegant and multifunctional. For events, it is very modern yet light, and can be branded. It starts from a simple basic model and has infinite possibilities. There are no limits to creativity.

Hersteller
Manufacturer
Inflate UK
London E8 3SB UK

Design
Designer
Werksdesign
In-house design
Nick Crosbie

Vertrieb
Distributor
Pulsmacher GmbH
D-71636 Ludwigsburg

Inflatables sind aufblasbare Raumelemente für weltweite Events sowohl im Innen- als auch Außenbereich. Das sind leichte, einfach aufzubauende, aufblasbare Räume und Architekturen, die teilweise in weniger als zehn Minuten realisiert werden können und aus dem Nichts zu entstehen scheinen. Damit eignen sie sich auch für außergewöhnliche und erschwingliche Messestände, die schnell und einfach auf- und abgebaut werden können.

Es gibt sie in vielfältiger Gestalt und in innovativem Design, kombiniert mit individuellem Branding und einer effektvollen Umgebungsbeleuchtung. Komfort bieten geräuscharme Lüfter und leistungsfähige Heiz- oder Kühlmodule. Den Anwendungsbeispielen sind keine Grenzen gesetzt: von leichten, freistehenden Airwalls, die etwa als Raumteiler fungieren über Chill-Out-Lounges, Meeting Space oder stylische Bar bis hin zu Sonderanfertigungen mit Brandings.

Die Vorteile des leichten Transports und des schnellen Aufbaus sind ideal für den temporären Einsatz auf Events. So werden Zeit und Kosten eingespart.

Das neueste Modell ist Boxer. Er entwickelt sich aus einem quadratischen Grundriss heraus, der sich nach einer Seite öffnet und aus dem Quadrat ausbricht. In dieser Gestalt erinnert er an ein Fort. Nach oben ist er offen, wobei die Wände sich nach innen biegen. Boxer ist für Innenräume konzipiert – er bietet einen Raum im Raum – und ist in verschiedenen Größen erhältlich.

Inflatables are room structures for global events, for use indoors and out. They are light, easy-to-install, inflatable rooms and architectures, some of which can be put up within ten minutes, and seem to materialize out of nowhere. This makes them suitable for unusual yet affordable trade fair stands, which can easily be assembled and knocked down again.

They are available in all kinds of shapes and in an innovative design, combined with individual branding and dramatic ambient lighting. They can be made more comfortable with low-noise fans and high-performance heating and cooling modules. There are no limits to where they can be used: from light, free-standing air-walls that act as partitions, to chill-out lounges, meeting spaces or stylish bars, to customized branded versions.

The advantages of light transport weight and fast installation are ideal for temporary use at events. This saves time and money.

Boxer is the latest model. Its basic form in a square that is open on one side. Its shape is reminiscent of a fort. It has no ceiling, but its walls bend inwards. Boxer is designed for interiors – offering a room within a room – and is available in various sizes.

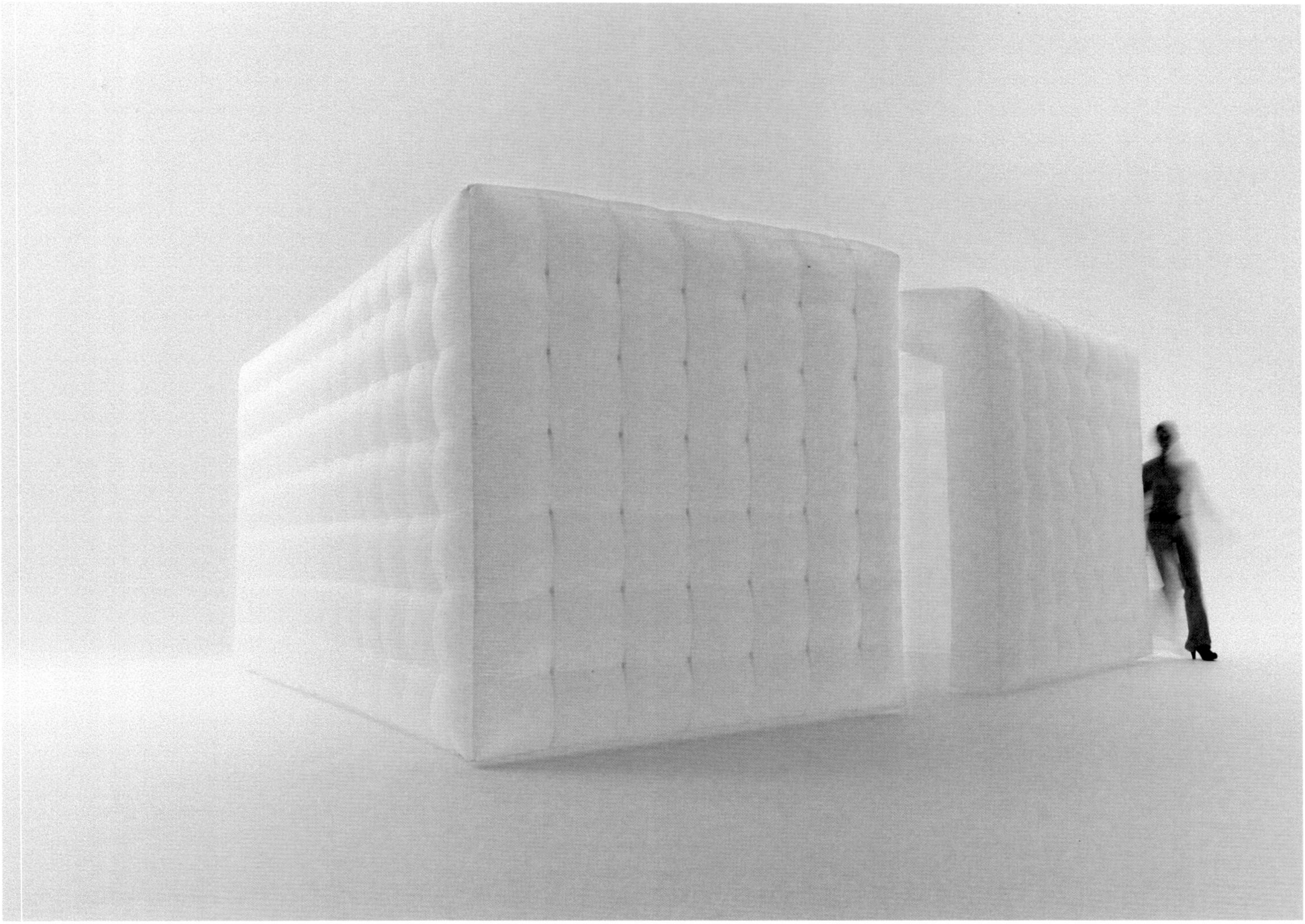

Ottobox
Besprechungsbox auf Rollen

Meeting box on casters

Jury
Judges panel

Public Design
Silber

Public design
Silver

Eine konsequente Weiterentwicklung der Officeboxen von Burkhardt Leitner, sehr minimalistisch. Alles wird flächenbündig eingelassen, sehr ruhig bei der Vielzahl der Elemente. Intelligente Details, einfache Befestigung. Leichter als das bisherige, dadurch auch ruhiger, klarer in der Linienführung. Über jedes Detail wurde genau nachgedacht, auch über Schallschutz. Es gibt eine optionale Belüftungsmöglichkeit, die eingebaut werden kann und sehr leise ist. Schöne Details.

A consistent further development of the Burkhardt Leitner office boxes, very minimalistic. Everything is flush mounted, nicely unruffled given the many elements it has. Intelligent details, simple fastening. Lighter than the previous models, and thus also calmer and clearer in its lines. Precise thought has been given to every detail, also sound insulation. There is an optional fan that can be installed and is very quiet. Attractive little details.

Hersteller
Manufacturer
Burkhardt Leitner
constructiv GmbH & Co. KG
D-70565 Stuttgart

Design
Designer
Werksdesign
In-house design
Team Burkhardt Leitner
constructiv

Vertrieb
Distributor
Burkhardt Leitner
constructiv GmbH & Co. KG
D-70565 Stuttgart

Die Ottobox bringt Bewegung in die Architektur, denn die Systemneuheit Constructiv Otto hat die Form einer Meetingbox, aber auf Rollen: einfach kurz anschieben und den Ort wechseln.

Trotz Rollen besitzt sie eine bodengeführte Glasschiebetüre – eine Neuheit bei den Officeboxen.

Im ersten Schritt ist die Ottobox eine Ergänzung zu den Officeboxen aus Constructiv Pon und Pila Petite. Auch sie lässt sich in allen Geschäftsbereichen einsetzen: Office, Showroom, Messe, Ausstellung. Sie greift dabei die bestehende Architektur nicht an, sondern hebt sich als autonome und flexible Raum-in-Raum-Lösung hervor.

Das System Constructiv Otto zeigt sich in einer geschlossenen Optik. Das Profil mit integriertem Knoten verfügt über eine hohe Biegesteifigkeit bei kleiner Dimension – wenige Bauteile, rascher Aufbau und Möglichkeiten für Befestigungen aller Art.

Die Ottobox ist in verschiedenen Größen erhältlich, serienmäßige Rastermaße sind dabei: 2,4 x 2,4 x 2,4 Meter oder 3,6 x 3,6 x 2,4 Meter. Das Corporate Design eines Unternehmens lässt sich bei der Box integrieren und kann sich in Farbe, Grafik und Typografie durch ganze Raumlandschaften ziehen. Die hohe Modularität der Ottobox ist ebenfalls für temporäre Werbeaktionen – Bedruckung von Glas, Segeln, Bannern et cetera – geeignet. Und es können auch individuelle Licht-, Belüftungs- und Akustiklösungen einfach und schnell integriert werden.

Ottobox brings mobility to architecture. The new Constructiv Otto system is shaped like a meeting box on casters. To change venue, a brief push is all that is needed.

Despite its casters, it has a floor-guided sliding glass door – a first among office boxes.

At first glance, Ottobox is a supplement to the Constructiv Pon and Pila Petite series. Like them, it can be used in any business setting: office, showroom, trade fair, exhibition. It does not impinge on the existing architecture, but stands out as an autonomous and flexible room-in-room solution.

The appearance of the Constructiv Otto system speaks for itself. Its profile with integrated nodes has high bending strength despite its small size – few components, rapid assembly and possibilities for attachments of all kinds.

Ottobox is available in various sizes. The standard grid sizes are: 2.4 x 2.4 x 2.4 metres or 3.6 x 3.6 x 2.4 metres. A company's corporate design can be integrated into the box, with colour, graphics and typography repeated across entire series of boxes. Ottobox's extremely modular characteristics make it also suitable for temporary advertising campaigns – with printed glass, sails, banners, and the like. And individual lighting, ventilation, and acoustic solutions can also be integrated simply and quickly.

Busch-iceLight®
LED-Orientierungslicht

LED orientation light

Jury
Judges panel

Public Design
Silber

Public design
Silver

Ein kleines Produkt, aber mit sehr vielfältigen Einsatzmöglichkeiten als Ambiente- oder als Orientierungslicht an Treppenaufgängen sowohl im Privatbereich als auch im halböffentlichen und öffentlichen Bereich. Von der Ausrichtung her sehr modern, quadratisch, kantig, aber mit stilistischen Effekten, die das Orientierungslicht emotionaler machen, oder als Piktogramm. Von daher eine sehr schöne gesamtheitliche, moderne Lösung, die sich auch über Standarddosen problemlos vom Elektrofachhandel nachträglich einbauen lässt.

A small product with a wide variety of possible uses. It can provide ambient light or orientation where there are stairs, for example. It is suitable for the private sphere, for the semi-public sphere or for business premises. Its basic design is very modern, with its square, straight-lined shape. But there are also stylistic elements that make the orientation light more emotional, for example when pictograms are used. This makes it a very appealing and very modern integrated solution. It can also be easily retrofitted onto standard sockets from an electrical supplier.

Busch iceLight® combines extremely tasteful design with groundbreaking LED technology. The light can be used in many different ways.

It comprises high-quality metal inserts, which are available in various grid sizes and designs. The light system comes in three flush-mounted variants: for walls and ceilings, and as an illuminated sign to help people find their way. This allows rooms to be set up individually with appropriate lighting, depending on the room's purpose.

The orientation light fulfils three tasks: rapid orientation, clear information and characterization. This is achieved by means of high quality metal blinds. To make them especially easy to read, the symbols are negatives. There are 30 of them to choose from – for nearly every requirement in the business and semi-public sphere.

Hersteller
Manufacturer
Busch-Jaeger Elektro GmbH
D-58513 Lüdenscheid

Design
Designer
Hadi Teherani
D-20457 Hamburg

Vertrieb
Distributor
Busch-Jaeger Elektro GmbH
D-58513 Lüdenscheid

Busch-iceLight® verbindet hohe Ästhetik mit zukunftsweisender LED-Technologie. Das Orientierungslicht ist vielseitig einsetzbar.

Es besteht aus hochwertigen Metalleinlegern, die in unterschiedlichen Rastern und Dekoren erhältlich sind. Das Lichtsystem besteht aus drei Varianten der Unterputzdose: für Wand und Decke und als Lichtzeichen zur Orientierung. Damit können Räume ganz individuell je nach Zweck und Bestimmung mit dem passenden Licht gestaltet werden.

Das Orientierungslicht erfüllt drei Aufgaben: schnelle Orientierung, klare Information und Differenzierung. Dafür sorgen hochwertige Designblenden aus Echtmetall. Damit man sie besonders gut lesen kann, sind die Symbole negativ dargestellt. 30 Varianten sind davon erhältlich – für nahezu alle Anforderungen im geschäftlichen und halböffentlichen Bereich.

Hersteller
Manufacturer
Recaro Aircraft Seating
GmbH & Co. KG
D-74523 Schwäbisch Hall

Design
Designer
Werksdesign
In-house design

Vertrieb
Distributor
Recaro Aircraft Seating
GmbH & Co. KG
D-74523 Schwäbisch Hall

BASIC LINE 3520 RECARO

Basic Line 3520
Economy-Class-Flugzeugsitz

Economy-class aircraft seat

Transport, Verkehr
Gold

Transport, traffic
Gold

Basic Line 3520
Economy-Class-Flugzeugsitz

Economy-class aircraft seat

Transport, Verkehr
Gold

Transport, traffic
Gold

Jury
Judges panel

So muss ein Flugzeugsitz sein: leicht, minimaler Materialeinsatz und bequem. Durch die schlanke, reduzierte Bauweise gewinnt man drei Zoll Raum. Kleine, feine Details sind ein Kleiderhaken an der richtigen Stelle oder die Positionierung der Magazintasche. Entscheidend ist auch die Gewichtsersparnis: 30 Prozent leichter als die meisten Flugzeugsitze. Gut gestaltet: Der Bezug hebt sich in Farbe, Textil, Schnitt, Verarbeitung und Design vom amerikanischen Kunstleder-Muff ab.

Just how an aircraft seat should be: light, minimum use of materials, comfortable. The slim, reduced design results in a gain of three inches more space. Good attention to detail, such as a coat hook in the right place, or the positioning of the pocket for brochures. Apart from the added space, the weight saving is decisive. Thirty percent lighter than most of the aircraft seats in use. Well designed: the seat covering's colour, textiles and cut, as well as the way it has been worked and designed, make it stand out from the boring American leatherette.

Hersteller
Manufacturer
Recaro Aircraft Seating
GmbH & Co. KG
D-74523 Schwäbisch Hall

Design
Designer
Werksdesign
In-house design

Vertrieb
Distributor
Recaro Aircraft Seating
GmbH & Co. KG
D-74523 Schwäbisch Hall

Ein leichter Flugzeugsitz mit hohem Komfort und modernem Design – ein Widerspruch? Nicht bei Basic Line 3520. Der Economy-Class-Sitz kombiniert geringes Gewicht, Premium-Qualität und attraktive Gestaltung mit neuen ergonomischen Erkenntnissen.

Im Vergleich zum Vorgängermodell wurde der Platz für den Fluggast um mehr als drei Zoll (7,6 Zentimeter) dank der dünnen Rückenlehne sowie der Verschiebung der Literaturtasche vergrößert. Der neu konstruierte Sitzboden bietet mehr Komfort als das bisherige Sitzkissen, Druckstellen an den Oberschenkeln werden reduziert.

Der Rahmen der Rückenlehne hat eine ergonomisch richtige Kontur und besteht aus neuen Materialien. Statt der üblichen Schäume bildet eine innovative Netzbespannung den Kern der Lehne. Der Sitz lässt sich zudem verstellen. Ein Sitzplatz wiegt nur etwas mehr als zehn Kilo und ist damit 30 Prozent leichter als das Vorgängermodell. Erreicht wurde das durch Hightech-Materialien und neue intelligente Konstruktionen.

Dank des vergrößerten Living Space können Airlines den Sitzabstand verringern und dadurch die Sitzanzahl um fünf bis zehn Prozent erhöhen. Dennoch bleibt dem Passagier mehr Raum als zuvor. Und trotzdem ist das Gesamtgewicht immer noch geringer als das der alten Sitze. Der Erstkunde beispielsweise spart rund 4,3 Prozent Kerosin ein und senkt so die CO_2-Emission.

Basic Line 3520 beweist, dass die schwierige Balance zwischen Kraftstoffeinsparung, ergonomisch optimiertem Sitzen und Premium-Qualität möglich ist.

A light, slim aircraft seat that is extremely comfortable and modern in design – a contradiction in terms? Not if it is the Basic Line 3520. This economy-class seat combines low weight, premium quality and attractive design with new ergonomic data.

Compared with its predecessor, the amount of space available for passengers has been increased by more than three inches (7.6 centimetres), thanks to the thin back rest and the relocation of the pocket for brochures. The newly designed seat bottom is more comfortable than the cushions that have been used so far, and there are fewer pressure points on the thighs.

The shape of the back-rest frame is ergonomically correct, and is made of new materials. Instead of the usual foamed material, an innovative mesh upholstery forms the core of the back rest. The seat can also be reclined. One seat weighs only slightly more than ten kilos, 30 percent lighter than its predecessor. This has been achieved thanks to high-tech materials and new, smart design.

Thanks to this increased space, airlines can reduce the space between seats, thus increasing the number of seats by 5 to 10 percent. Nonetheless, passengers have more space than previously, thanks to the innovative seat design. Moreover, overall weight is still less than that of the old seats. Customers can save roughly 4.3 percent fuel, and in this way reduce their CO_2 emissions.

Basic Line 3520 proves that it is possible to achieve the difficult balance between fuel saving, ergonomically optimized seats and premium quality.

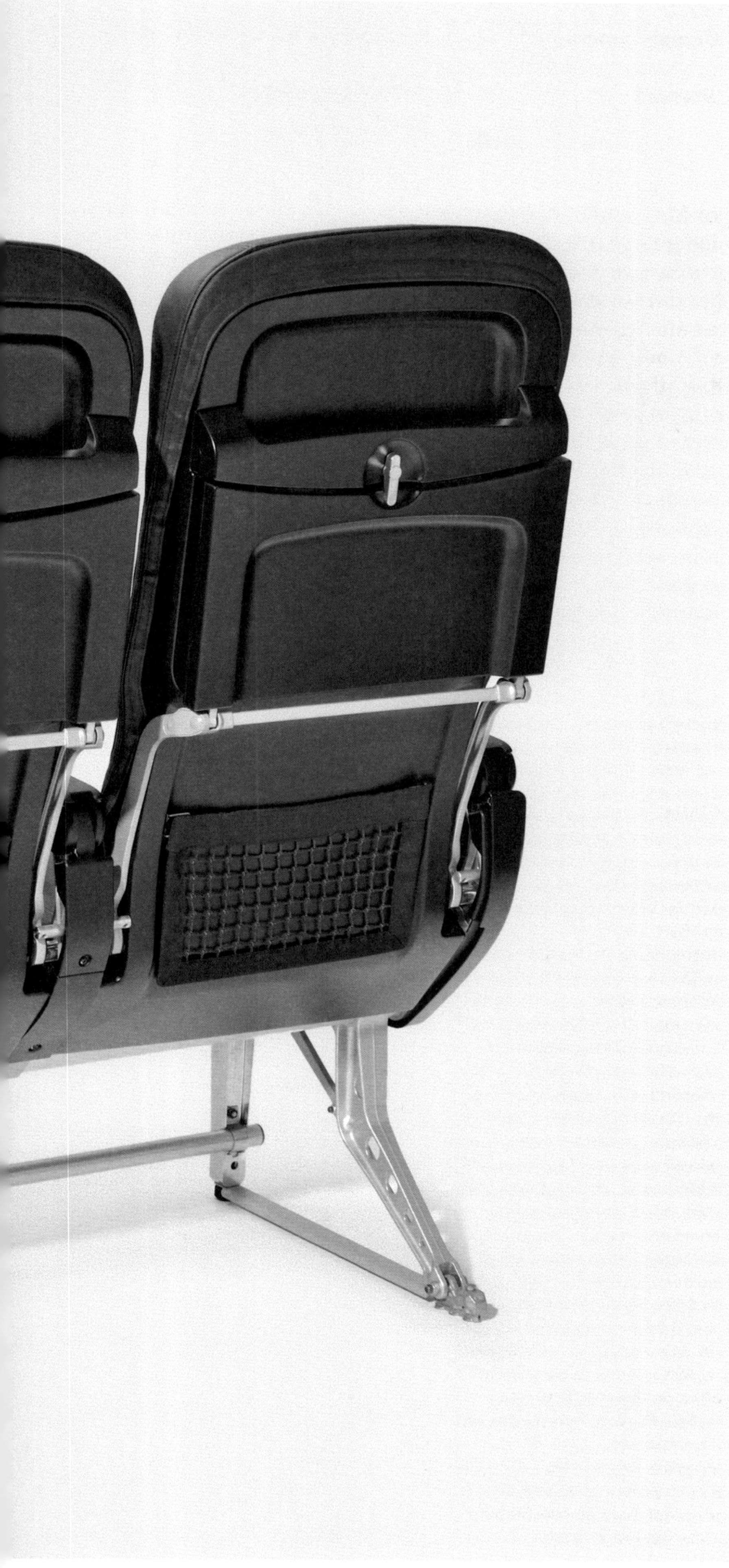

Tigis Ergo Rettungstransportwagen

Ambulance

Transport, Verkehr
Silber

Transport, traffic
Silver

Jury
Judges panel

Das Interieur ist eine hervorragend gestaltete Lösung, die durch viele sehr innovative Details und vorbildliche Rationalität besticht. Über Handlungsabläufe der Rettungssanitäter wurde intensiv nachgedacht und innovative Lichtsysteme, Informationsdisplays sowie ein neues Ablagemanagement geschaffen. Und es wurde an den Patienten gedacht. Mit der Beleuchtung kann man auf seine Befindlichkeit positiv Einfluss nehmen, was hier erstmalig in dieser Art umgesetzt wurde. Die Atmosphäre im Raum ist freundlich und nicht bedrückend.

The interior is an excellently designed solution, impressive for its many very innovative details and exemplary rationality. A lot of thought was given to the sequences of actions performed by paramedics, and this resulted in the creation of innovative light systems and information displays, as well as a new filing management system. And thought was given to the patient as well. Lighting can have a positive effect on patients' mood. This is the first time something like this has been done. The atmosphere inside the ambulance is friendly, not depressing.

Hersteller
Manufacturer
Ambulanz Mobile
GmbH & Co. KG
D-39218 Schönebeck

Design
Designer
Uwe Thamm
D-74595 Langenburg

Vertrieb
Distributor
Ambulanz Mobile
GmbH & Co. KG
D-39218 Schönebeck

Rettungsfahrzeuge baute man bisher immer nach dem gleichen Muster und mit den gleichen Möbeln. Mit dem Tigis Ergo wurde nun alles neu in Frage gestellt. Durch eine in dieser Branche einzigartige Herangehensweise schuf ein interdisziplinäres Entwicklungsteam einen progressiven Rettungswagen, der sich barrierelos bedienen lässt. Dabei flossen auch die gesammelten Erfahrungen der Rettungsassistenten mit ein.

Das Ergebnis ist ein auf Patient und Rettungspersonal abgestimmtes Interieur, das ergonomische Arbeitsbedingungen bietet, Ressourcen einspart und die Sicherheit erhöht. Es ist durch die sorgfältige Verarbeitung qualitativ hochwertig und ökologisch wegen der recycelfähigen ABS-Innenverkleidung.

Das Interieur wirkt aufgeräumt, jedoch nicht kalt oder steril sondern freundlich und frisch. Dafür sorgen auch die Staufächer, die in verschiedenen Farben zu haben sind, und die innovative Ambientebeleuchtung. Das Licht wird indirekt abgegeben. Die dadurch entstehende behagliche Raumstimmung kann je nach Zustand und Empfinden des Patienten in Farbe sowie Helligkeit variiert werden, was Ruhe und Geborgenheit vermittelt. Das sind ideale Bedingungen, um während der Fahrt zum Krankenhaus stabilisierend auf den Patienten einwirken zu können.

Mit dem neuen Bedien- und Informationssystem Touch control steuert der Rettungssanitäter einfach und intuitiv – besonders wichtig gerade auch in turbulenten Situationen – durch Berühren des Bildschirms Beleuchtung, Klimatisierung und Tragentisch.

Up to now, ambulances have always been built according to the same pattern, with the same furnishings. When designing the Tigis Ergo, these conventions were critically scrutinized. Using an approach that is unprecedented in this industry, an interdisciplinary team created a progressive ambulance that presents users with no barriers. The collected experience of ambulance personnel found its way into the design.

The result is an interior geared to the needs of patient and paramedics, offering ergonomic working conditions, saving resources, and increasing safety. Its careful workmanship lends it a very high quality. It is also eco-friendly, because it uses recyclable ABS for its interior trim.

The interior is tidy without being cold or sterile. In fact, the impression it makes is one of friendliness and freshness. This impression is underscored by the storage compartments, which are available in various colours, and by the innovative ambient lighting. The interior is illuminated indirectly. The cosy atmosphere this creates can be varied in colour and brightness depending on the condition and wishes of the patient, giving a sense of tranquillity and security. These are ideal conditions for stabilizing a patient on the way to the hospital.

With its new touch-control operating and information system, paramedics can control lighting, air-conditioning and stretcher platform simply and intuitively, which is especially important in hectic situations.

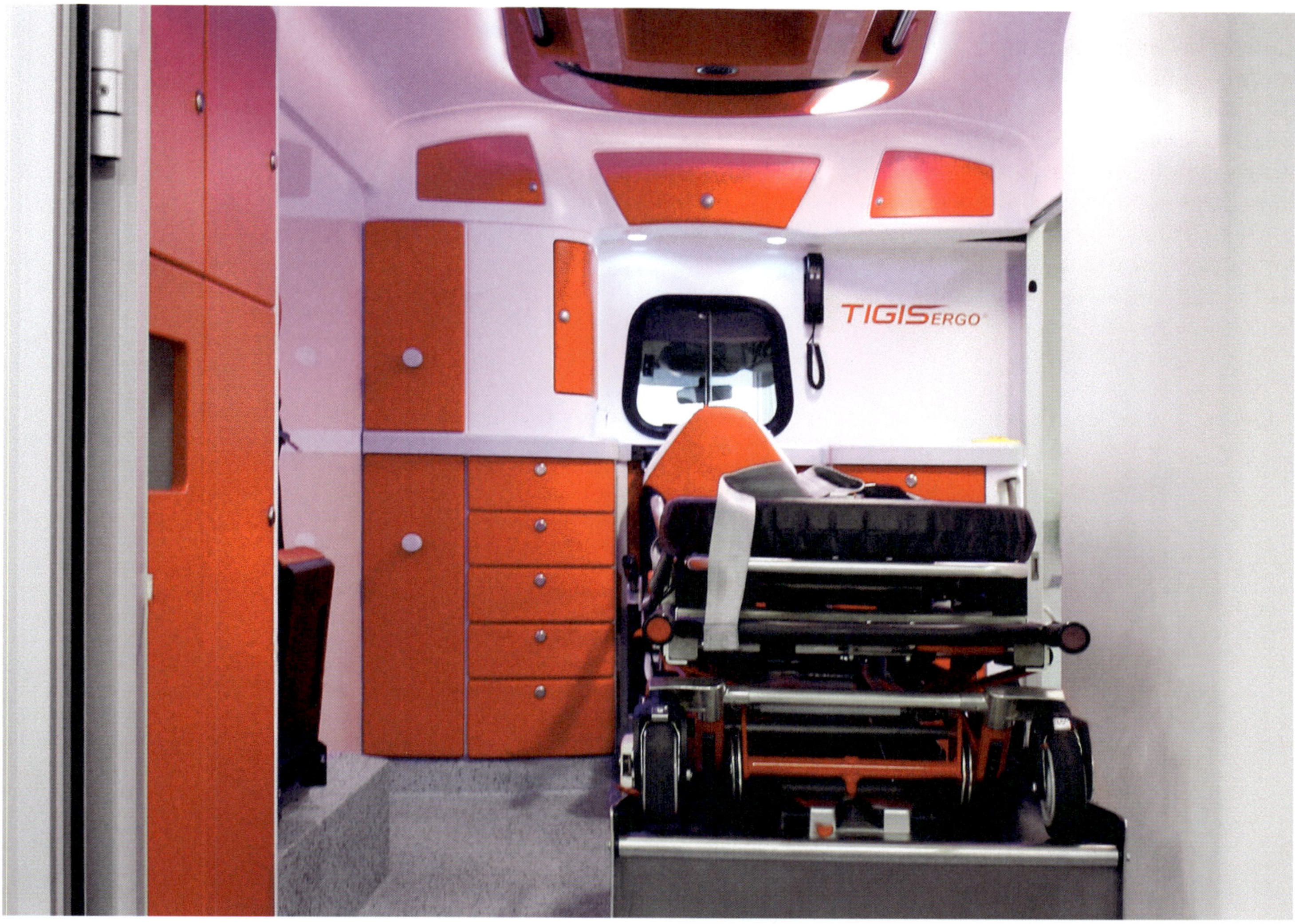
TIGISERGO

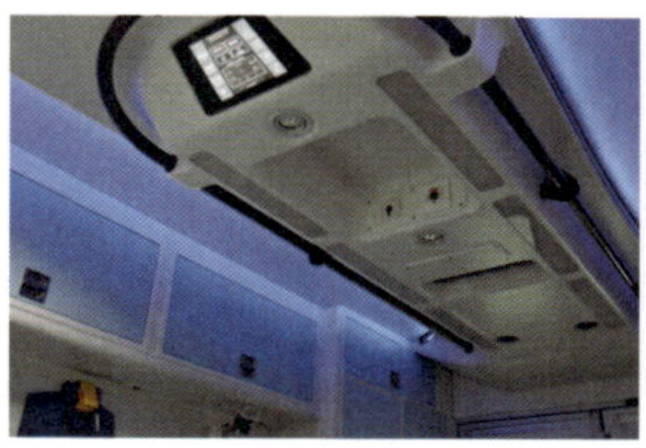

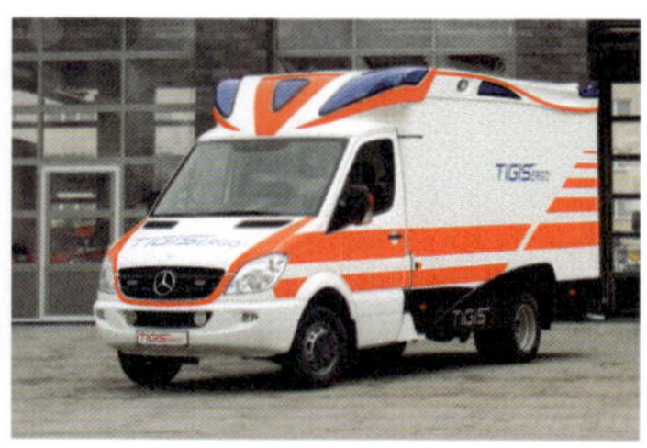

RUDmatic Hybrid Schneekette

Snow chain

Transport, Verkehr
Silber

Transport, traffic
Silver

Jury
Judges panel

Die Schneekette überzeugt uns, weil sie sehr einfach zu bedienen ist. Sie ist sowohl leicht zu montieren als auch zu demontieren und auch für Menschen mit wenig Kraft gut handhabbar. Es überzeugt uns, dass die Schneekette auch auf Eis greift. Und sie erfüllt strenge Normen, besonders wichtig in Ländern, in denen nur Ketten und keine Textilvarianten zugelassen sind.

We like the snow chain because it is easy to use. It is easy both to fit and to remove, even for people who are not very strong. We are impressed by the way it even grips on ice. And it complies with strict standards, in countries where only snow chains are allowed, for example.

Hersteller
Manufacturer
RUD Ketten
Rieger & Dietz GmbH & Co. KG
D-73432 Aalen

Design
Designer
Werksdesign
In-house design

Die neuartige Hybrid-Schneekette sorgt für Montagefreundlichkeit und höchsten Fahrkomfort. Sie besteht aus einem Laufnetz aus Stahl und Seitenpartien aus Textilmaterial. Durch diese Kombination ist ein Quantensprung in der Schneekettentechnologie gelungen. Sie macht sie besonders montagefreundlich und sorgt dafür, dass die Ketten ein geringes Gewicht haben.

Ohne Haken und Ösen wird die Schneekette einfach über das Laufrad gestülpt. Ein Gummi mit Kunststoffhülsen erleichtert die Montage und Demontage. Auch auf die Laufrichtung muss beim Aufziehen nicht mehr geachtet werden – RUDmatic Hybrid lässt sich problemlos links und rechts und je nachdem, welche Achse angetrieben wird, vorne oder hinten montieren. Damit erfolgt die Montage intuitiv und garantiert ein Höchstmaß an Bedienungsfreundlichkeit.

Das Laufnetz aus Feinkornstahl, das wie ein Kettengürtel gestaltet ist, sorgt auf schnee- oder eisbedeckter Fahrbahn für eine optimale Traktions- und Bremswirkung. Dank der textilen Seitenteile sind auch empfindliche Leichtmetallfelgen ideal geschützt, da sie nicht mit den Metallteilen der Kette in Berührung kommen.

Außerdem erfüllt sie im Unterschied zu den reinen Textilvarianten die strenge Norm für Österreich und ist somit auch dort zugelassen, wo ausdrücklich Schneekettenpflicht herrscht.

This innovative hybrid snow chain is easy to assemble and makes for a comfortable drive. It comprises a steel chain mesh and textile side parts. The combination brings about a quantum leap in snow-chain technology. It makes the snow chain especially easy to fit, and keeps the weight of the snow chain down.

The snow chain is simply pulled over the drive wheel, with no need for hooks and eyes. An elastic band with plastic sleeves makes fitting and removal easier. Moreover, attention no longer has to be paid to the direction of travel when fitting the chain. RUDmatic Hybrid can be fitted easily to the nearside and offside, as well as to the front and rear, depending on which axle is the driven one. This makes fitting the chain intuitive, and guarantees extreme ease of use.

The chain mesh of fine-grained steel, which is shaped like a chain belt, ensures optimum traction and braking on snow-covered or icy roads. The textile side panels mean that even sensitive alloy wheel rims are ideally protected, since they do not come into contact with the metal parts of the chain.

In addition, unlike its fully textile version, this snow chain fulfils the strict Austrian road standards, and so is also approved for use wherever snow chains are mandatory.

RUD matic Hybrid
RUD matic Hybrid
RUD matic Hybrid
RUD
RUD®
RUD matic Hybrid
RUD matic Hybrid
RUD matic Hybrid

Hersteller
Manufacturer
Daimler AG
D-71059 Sindelfingen

Design
Designer
Werksdesign
In-house design

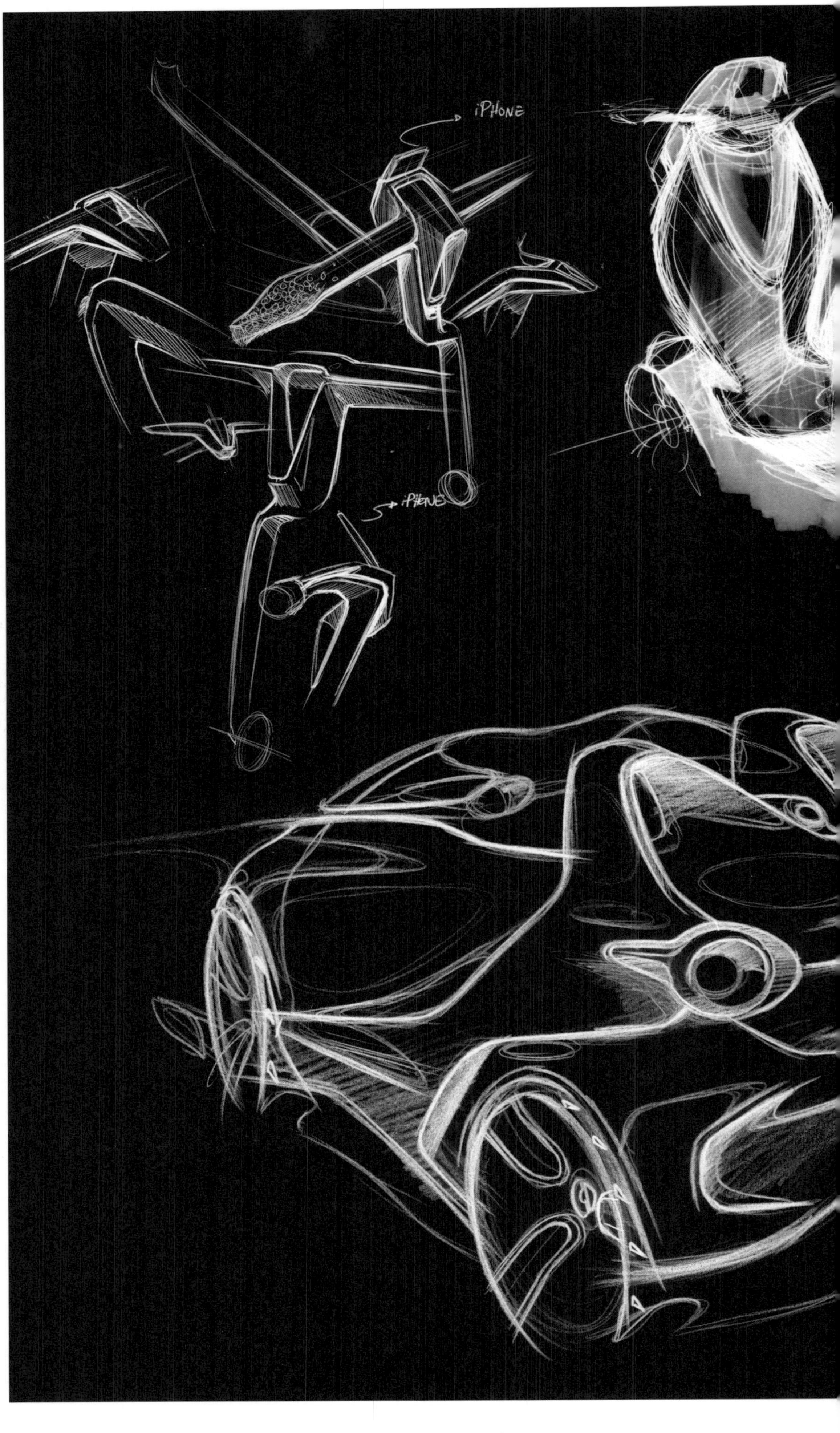

SMART STUDIE DAIMLER

E-mobility
Konzeptstudie

Konzept, Nachhaltigkeit
Gold

Design study

Design studies for sustainability
Gold

E-mobility
Konzeptstudie

Design study

Konzept, Nachhaltigkeit
Gold

Design studies for sustainability
Gold

Jury
Judges panel

Die Konzeptstudie ist ein Zeichen für den richtigen Weg: Mobilität im Stadtverkehr und alternative Antriebstechnologien. Wir würden uns von vielen Autoherstellern wünschen, dass sie über solche Wege nachdenken. Über allem steht, dass der Umweltgedanke sich mit dem Hedonistischen verbindet, was bei allen drei Projekten gut umgesetzt ist. Das ist die richtige Richtung für ein Unternehmen, das Autos herstellt und sich nicht nur über vierrädrige, sondern auch über zweirädrige Fahrzeuge Gedanken macht.

The design study is a step in the right direction: mobility in urban traffic and alternative drive technologies. We would like to see lots of carmakers thinking about such steps. The overriding concern is one of combining environmental considerations with hedonism. This has been done well in all three projects. This is the right direction for a company that makes cars and thinks not only about four-wheeled vehicles, but also about two-wheeled ones.

Hersteller
Manufacturer
Daimler AG
D-71059 Sindelfingen

Design
Designer
Werksdesign
In-house design

Smart präsentiert eine unkonventionelle Studie, die zeitgemäße urbane Mobilität mit dem Smart-typischen Spaßfaktor verbindet.

Bei Smart hat das Zeitalter der Elektromobilität bereits begonnen. Seit Ende 2009 wird mit dem Smart fortwo electric drive schon die zweite Generation eines elektrisch angetriebenen Smart an Kunden in 18 Ländern ausgeliefert. Dieses Elektroauto erfreut sich einer großen Nachfrage. Damit verfügt Smart über umfassendes Kundenfeedback, das in die Studie eingeflossen ist.

Nun zeigt die zukunftsweisende Konzeptstudie Lösungen für intelligente urbane Mobilität und damit einen besonders spannenden Ausblick auf kommende Generationen von Elektrofahrzeugen.

Die Studie beweist, wie elektrisch fahren in allen Variationen möglich sein kann und präsentiert eine interessante strombetriebene Modellfamilie. So wird neben dem Auto, dem Smart forspeed, ein Elektroroller, der Smart escooter, und ein Elektrofahrrad, das Smart ebike, vorgestellt. Der Automobilhersteller will damit die Innenstädte weiter von Emissionen entlasten helfen. Hier wird Verantwortung und Emotion auf eine neue, lustvolle Weise kombiniert. Der Spaßfaktor wird dabei großgeschrieben.

Allen drei Fahrzeugen gemeinsam sind Smart-typische Designmerkmale – wie innovative Formgebung –, die einfache Handhabung sowie die Integration von Smartphones als Steuer- und Kommunikationszentrale. Die weiterentwickelte smart drive kit App übernimmt dabei vielfältige Funktionen.

Smart presents an unconventional study that combines contemporary urban mobility with the fun factor that is typical of smart.

At smart, the age of electromobility has already begun. In the form of the smart fortwo electric drive, the second generation of an electric vehicle has been available for customers in 18 countries since the end of 2009. This electric car is in great demand. This means that smart has a lot of customer feedback, and was able to feed it into this study.

This pioneering design study presents solutions for smart urban mobility, and thus an especially exciting outlook on the future generations of electric vehicles.

The study proves that electric driving is possible in every variation, and presents an interesting electrically driven model family. Apart from a car, the smart forspeed, it presents an electric scooter, the smart e-scooter, and an electric bicycle, the smart e-bike. The carmaker's intention is to help further reduce emissions in inner cities. Responsibility and emotion are combined in a new, sensual way. Above all, the emphasis is on fun.

All three electric vehicles feature design characteristics that are typical of smart, such as innovative and daring lines. They are also easy to handle and allow smartphones to be integrated as a control and communications centre. The modified smart drive kit app assumes a variety of functions here.

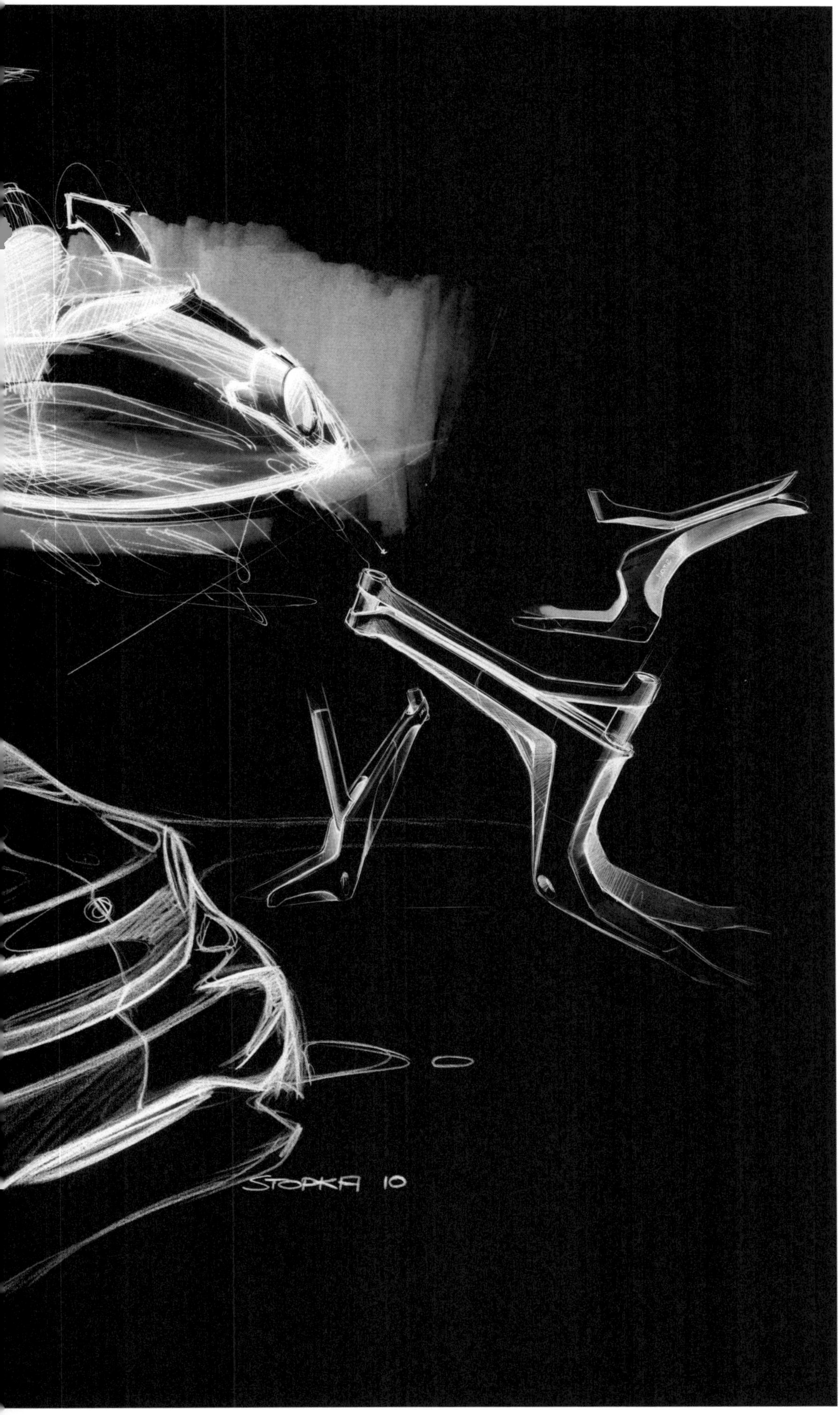
STOPKA 10

Jury
Judges panel

Smart forspeed
Konzeptfahrzeug

Vehicle study

Konzept, Nachhaltigkeit
Silber

Design studies for sustainability
Silver

Das Konzeptfahrzeug verbindet viele schöne und innovative Ideen. Zum Beispiel findet man das Smart-Zeichen am Automobil in den einzelnen Elementen wieder, ohne dass es explizit auf die Karosserie appliziert ist. Die Lichter sind hervorragend ausgearbeitet, zum Beispiel die integrierten Blinker – ein sehr feines und besonderes Detail. Das Auto besticht durch einen Retro-Futurismus: 1955 – Peter Krauss – Gardasee, aber auf 2022 interpretiert. Das halten wir für eine tolle Geschichte, damit verbindet sich Emotionalität.

The vehicle study combines many attractive and innovative ideas. For example, the smart logo is to be found in the car's various elements, yet it is not explicitly displayed on the body. The lights have been excellently done – the way the direction indicator is integrated in them, for example – a very nice, special detail. The car's retro-futurism is captivating: it's like the automotive freedom of the fifties, but reinterpreted for the year 2022. A super story, which gives it its emotional appeal.

Hersteller
Manufacturer
Daimler AG
D-71059 Sindelfingen

Design
Designer
Werksdesign
In-house design

Ein unvergleichliches Frischluftvergnügen bietet der Zweisitzer ohne Dach und Seitenscheiben – statt Frontscheibe besitzt er einen flachen Windabweiser im Stil klassischer Propellerflugzeuge. Der antrittsstarke, emissionsfreie Elektroantrieb garantiert den Fahrspaß.

So beschleunigt der Smart forspeed von 0 auf 60 Stundenkilometer in nur 5,5 Sekunden und ist bis zu 120 Stundenkilometer schnell. Aktiviert man über eine Taste in der Mittelkonsole die Boost-Funktion, bringt er kurzzeitig fünf Kilowatt mehr Leistung – ideal, um beispielsweise auf der Stadtautobahn zu überholen. Nach 135 Kilometern – so lange reicht eine Ladung – wird die Lithium-Ionen-Batterie einfach an einer konventionellen 220-Volt-Steckdose aufgeladen.

Wetterschutz bietet eine Persenning, die über das Interieur gespannt wird. Auch ohne sie wird ein Platzregen nicht zum Problem, dafür sorgen Wasserablaufkanäle in Boden und Sitzflächen. Im Cockpit bietet ein Smartphone alle Features, die während einer Fahrt gebraucht werden: Telefonieren über Freisprechanlage, Navi und Car Finder.

Smart forspeed zeichnet sich durch einen dynamischen Auftritt aus. Auffällig ist das Design der Scheinwerfer. Exterieur und Interieur sind als formale Einheit gestaltet. So fließt die tridion-Sicherheitszelle hinter den Überrollbügeln nach vorne und bildet als tragendes Teil den Mitteltunnel. Runde und gespannte Formen, der Verzicht auf Fugen, glatte, grifflose Oberflächen sowie ein an den Flugzeugbau angelehntes Design senken den Luftwiderstand.

This two-seater without a roof and side windows offers incomparable open-air driving pleasure: instead of a windscreen, it has a flat wind deflector of the kind seen in classic propeller-driven aeroplanes. The speedy, emissions-free electric motor guarantees driving fun.

The Smart forspeed accelerates from 0 to 60 kilometres per hour in just 5.5 seconds, and has a top speed of 120 kph. Pressing a button in the central console activates a boost function, which temporarily increases motor power by another five kilowatts – which is ideal when overtaking on a dual carriageway, for example. After 135 kilometres – which is the limit for one battery charge – the lithium-ion battery is simply recharged from a conventional 220-volt power socket.

A tarpaulin spanned over the interior provides protection from the elements. Even without the tarpaulin, a sudden shower is no problem, as water drainage channels have been built into the vehicle floor and seat surfaces. In the cockpit, a smartphone offers all the features that are needed while driving: phoning via a hands-free kit, sat nav and car finder.

Smart forspeed stands out by virtue of its dynamic appearance. One striking feature is the design of the lights. The exterior and interior make up a formal unity. The tridion safety cell behind the rollover bars flows to the front, forming the central tunnel as a load-bearing member. Rounded, taut forms, the deliberate lacks of joints, smooth, handle-less surfaces, as well as a design based on aircraft construction lower air resistance.

Smart escooter
Zweirad-Studie

Two-wheeler study

Jury
Judges panel

Konzept, Nachhaltigkeit
Silber

Design studies for sustainability
Silver

Der Roller hat ein ausgesprochen gelungenes Grundkonzept: Man kann im Stadtverkehr mit einem Elektrofahrzeug fahren und es überall aufladen – das ist hervorragend. Dazu hat es extrem gut ausgearbeitete Details, wie die Leuchten und die markanten Schutzleisten, die gleichzeitig Blinker sind, der Stauraum – wichtig für Stadtverkehr und Einkäufe –, die Aufladestation, kleiner Wendekreis, der Soziussitz, der inklusive der Fußstützen ausgeklappt werden kann. Auch die Integration des Smartphone, das gleichzeitig Tachometer und Navi ist, ist sehr schön.

The basic concept behind the scooter is exceptionally attractive. You can use an electric vehicle in urban traffic and recharge it anywhere – that's excellent. And its attention to detail is excellent, such as the lights and the striking protective strips, which also serve as direction indicators, the stowage space – important for urban driving and shopping – the charging unit, the small turning radius, and the pillion seat that can be folded out, including foot rests. The way the smart phone has been integrated, so that it is also a speedometer and a sat nav, is also very well done.

Hersteller
Manufacturer
Daimler AG
D-71059 Sindelfingen

Design
Designer
Werksdesign
In-house design

Die innovative Zweirad-Studie überträgt die typischen Vorteile des E-Auto-Pioniers Smart fortwo electric drive konsequent auf zwei Räder: kompakte Abmessungen, Wendigkeit, hohe Sicherheitsstandards und markentypischer Fahrspaß, nicht zuletzt durch einen effizienten und emissionsfreien Antrieb. Zusätzlich zu seinen innovativen technischen Lösungen überzeugt der neue escooter durch sein eigenständiges und modernes Design sowie durch einfache Handhabung.

Für den Antrieb sorgt ein scheibenförmiger, sehr kompakter und leistungsfähiger Radnaben-Elektromotor im Hinterrad. Damit steht nun der Platz im Chassis, wo sonst der Motor untergebracht ist, für mehr Stauraum zur Verfügung. Das Smartphone übernimmt mit der weiterentwickelten smart drive kit App vielfältige Funktionen, wie Tachometer, Wegfahrsperre oder Reichweitenanzeige.

Beim Smart escooter gehen Funktionalität, Umweltverträglichkeit und Fahrspaß Hand in Hand. Mit ihm kann die Marke auch jüngere Menschen ohne Autoführerschein erreichen. Gerade im urbanen Verkehr ist der Elektroroller eine sinnvolle Ergänzung oder Alternative zum Auto. Neben vorbildlichem Komfort bietet er eine wegweisende Sicherheitsausstattung: Airbag, Antiblockiersystem ABS und einen Totwinkel-Assistenten.

Ein ausklappbarer Sitz und automatisch ausfahrende Fußstützen bieten Fahrspaß zu zweit. Benötigt man mehr Stauraum, so kann man den Soziussitz durch eine Gepäckbrücke ersetzen.

This innovative two-wheeler study systematically transfers the typical advantages of the electric-car pioneer smart fortwo electric drive to a two-wheel platform: compact dimensions, manoeuvrability, high safety standards, and the fun-to-ride features that are typical of the brand. Not least, its motor is efficient and emissions-free. In addition to its innovative technical solutions, the appeal of the new e-scooter lies in its unconventional and modern design and its simple handling.

It is driven by a disc-shaped, very compact and powerful electric motor positioned on the hub of the rear wheel. As the motor would otherwise be in the chassis, more room is now available there for stowing things away. Using its modified smart drive kit app, a smartphone assumes a variety of functions, including speedometer, immobilizer and battery-charge display.

In the smart e-scooter, functionality, eco-friendliness and fun go hand in hand. It allows the brand to reach younger people without a driving licence as well. Especially in urban traffic, this electric scooter makes sense as a complement or alternative to the car. Apart from exemplary comfort, it features groundbreaking safety equipment: airbag, ABS antilock braking system and a side view assist function.

A fold-out seat and automatically extending foot rests provide room for a second passenger. If the rider needs more stowage space, he can replace the pillion seat with a luggage rack.

Smart ebike
Elektrofahrrad-Studie

Konzept, Nachhaltigkeit
Silber

Smart ebike

Design studies for sustainability
Silver

Jury
Judges panel

Gestalterisch sehr ausgereiftes Produkt. Überzeugt uns durch viele schön ausgebildete Elemente, die ganz ungewöhnliche neue Strukturen suchen. Sieht leicht aus, erinnert eher an ein normales Fahrrad, weil der Antrieb hinten in der Nabe und der Akku unter der Rahmenverkleidung untergebracht ist. Vor allen Dingen ist er nicht additiv, sondern integriert, was bei den meisten Elektrofahrrädern nicht der Fall ist. Schön auch das integrierte Licht und charmante Details wie der Zahnriemen anstelle einer normalen Kette.

Design-wise, a very mature product. We like its many well designed elements, which call out for completely unconventional new structures. Looks light, is more reminiscent of a normal bike, because the drive is situated on the rear hub and the rechargeable battery is concealed under the frame housing. Above all, its design is integrated, not additive like in the case of most e-bikes. We also like the integrated light and the charming details like the drive belt instead of a normal chain.

Hersteller
Manufacturer
Daimler AG
D-71059 Sindelfingen

Design
Designer
Werksdesign
In-house design

Mit dem ebike präsentiert Smart eine weitere Idee für Elektromobilität auf zwei Rädern. Anders als der Roller wird der Motor hier nicht via Gasgriff gesteuert. Stattdessen liefert er die zusätzliche Leistung dann, wenn der Fahrer in die Pedale tritt. Der Muskelkraft-Elektromotor-Hybridantrieb stellt vier Leistungsstufen bereit.

Auch optisch und konzeptionell setzt das Hightech-Bike Maßstäbe, angefangen beim futuristischen Aussehen und den eigenständig gestalteten Leichtbaurädern bis hin zur technischen Ausstattung im Detail. Die Aufgaben von Kettenblatt und Kette übernehmen eine Riemenscheibe und ein Zahnriemen, der geräusch- und wartungsfrei arbeitet.

Beim ebike erfüllt das Smartphone mit der weiterentwickelten smart drive kit App vielfältige Funktionen, wie Tachometer, Navi, Pulsmesser, Anlasser und Wegfahrsperre.

Neben Eyecatcher-Design und innovativer Technik bietet das Elektro-Bike eine weitere Stärke: seine Vielseitigkeit. Im Alltagseinsatz – etwa auf dem Weg zur Schule, Uni oder Arbeit – schont der Elektromotor die Kondition des Fahrers: Steigungen verlieren dank »Elektro-Power« ihren Schrecken. Hinzu kommt die einfache Bedienung ohne konventionelle Gangschaltung. Gleichzeitig eignet sich das Hightech-Fahrrad für sportliches Fahren. Der ins Hinterrad integrierte E-Motor schiebt nicht unbegrenzt mit an, sondern wird bei 25 Stundenkilometern automatisch abgeregelt. Noch ein Vorteil: Beim Bremsen gewonnene Energie wird in elektrische Energie umgewandelt und im Akku, der sich unter der Rahmenverkleidung befindet, gespeichert.

With its e-bike, smart presents a further idea for two-wheeled mobility. Unlike the scooter, the e-bike's motor is not controlled by a throttle. Instead, it delivers extra power when the rider steps on the pedals. This hybrid of muscle power and electric drive provides four different levels of power.

Visually and conceptually as well, this high-tech bike sets standards, from its futuristic appearance and unconventionally designed lightweight wheels to the details of its technical equipment. Instead of a chain wheel and chain, there is now a pulley and drive belt, which works noiselessly and without any need for maintenance.

As with the e-scooter, a smartphone using a modified smart drive kit app assumes a variety of functions, including speedometer, sat nav, heart-rate monitor, starter and immobilizer.

Apart from its eye-catching design and innovative technology, the e-bike has a further special strength: its versatility. In everyday use – on the way to school, university or work – the electric motor keeps the rider from exerting himself. Thanks to electrical power, gradients are no longer anything to fear. Moreover, it is simple to use, without conventional gear shifts. At the same time, this high-tech bike is suitable for sporty riding. The electric motor integrated in the rear wheel does not provide unlimited support, but automatically cuts out when a speed of 25 kph is reached. And there is a further plus point: the energy generated during braking is converted into electricity and stored in the battery beneath the frame housing.

SportCONTACT

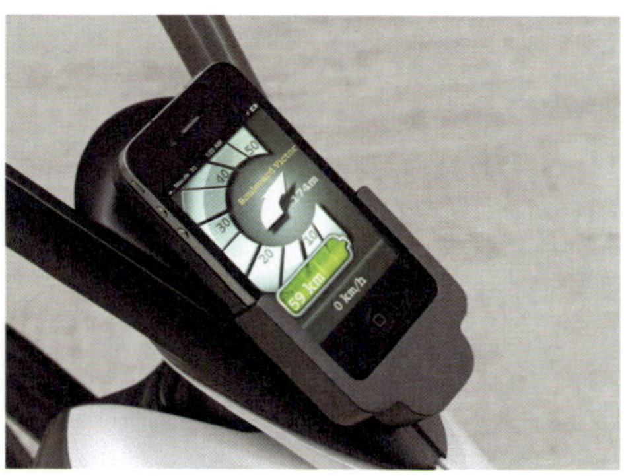

25 Jahre Mia Seeger Stiftung

25 years of the Mia Seeger Foundation

»Was ich nicht fördern möchte, ist dieses modische kurzlebige Zeug, wie es uns die Medien ständig als »Das Neue Deutsche Design« vorsetzen. Auch wenn es noch so spektakulär vorgestellt wird – damit wird dem Förderungsgedanken des gebrauchstüchtigen Designs kein Dienst erwiesen.«

»What I don't want to foster is the modish, short-lived kind of stuff that the media are constantly presenting as »new German design«. How ever spectacularly it may be presented – it will do the idea of promoting beneficial design no good.«

Mia Seeger, 1986

Am 19. September 1986 unterschrieb Mia Seeger die Urkunde, die zur Gründung der »Mia Seeger Stiftung – zur Förderung des Design« führte. Intensive Gespräche mit dem damaligen Präsidenten des Landesgewerbeamtes, Dr. Karl Reuss, der die Idee einer privaten Stiftung zur Förderung des Design-Nachwuchses begeistert aufgriff, führten dazu, dass bedeutende Zustifter gewonnen werden konnten, unter ihnen Daimler, die BW-Bank und Stihl. Die Anbindung der Stiftung an das Design Center Stuttgart und an die allgemeine Designförderung des Landes Baden-Württemberg war ihr ein besonderes Anliegen.

Seit 1988 wurden aus dem Kapitalertrag Preise und seit 1989 Stipendien der Mia Seeger Stiftung an junge Designer vergeben, die nach dem Urteil der jeweiligen Jury denjenigen zukommen, die sich, um »gebrauchstaugliches, sorgfältiges Design« bemüht haben oder im Rahmen eines Stipendiums »sinnvolle« Projekte umsetzen wollen. Eine spätere Satzungsänderung hat die Orientierung zur Bildung im Design hin festgeschrieben.

In den 90er Jahren bot die jährliche Designbörse im Haus der Wirtschaft den Rahmen für den Mia Seeger Preis. Seit gut zehn Jahren ist seine Präsentation in Katalog, Ausstellung und Preisverleihung eng mit dem »Internationalen Designpreis – Focus Open« verbunden. In fruchtbarer Weise versuchen sich Design Center Stuttgart und Mia Seeger Stiftung zum Nutzen der jungen Designer zu ergänzen.

Mia Seeger wurde am 9. Mai 1903 in Bad Cannstatt, heute mit Stuttgart vereint, geboren. Sie starb 88-jährig am 14. Mai 1991 in Stuttgart. Sie widmete ihre Lebensarbeit im weitesten Sinne dem, was wir Design nennen. Sie hatte die besten Lehrmeister, darunter Mies van der Rohe und Walter Gropius, sie hatte als Lernende teil an den wichtigsten

On 19 September 1986, Mia Seeger signed the deed that would lead to the establishment of the »Mia Seeger Foundation for the Promotion of Design«. As a result of intensive talks with the then president of the Office for the Promotion of Trade and Industry, Dr. Karl Reuss, who enthusiastically seized on the idea of a private foundation to promote young designers, it was possible to attract significant donors, including Daimler, BW-Bank and Stihl. Mia Seeger was especially concerned that the foundation should be affiliated to the Design Center Stuttgart and the general promotion of design by the state of Baden-Württemberg.

Since 1988, the yield from this capital has financed prizes, and since 1989 Mia Seeger Foundation grants, for young designers. They are given to young people who, in the eyes of the respective judges, have attempted to create »fit-for-purpose, careful design« or want to realize »meaningful« projects with the help of a grant. A subsequent amendment to the statutes stipulated an orientation towards design education.

In the 1990s, the annual design exchange in the »Haus der Wirtschaft« provided the setting for the Mia Seeger Prize. For a good ten years now, it has been closely tied with the »International Design Award – Focus Open«, sharing with it the same catalogue, exhibition and award ceremony. The Design Center Stuttgart and the Mia Seeger Foundation attempt to complement each other, for the benefit of young designers.

Mia Seeger
Stiftung

»Ich bin davon ausgegangen, dass eigentlich junge Leute einer Hilfe bedürfen (auch einer finanziellen), um ihren Blick zu erweitern. Mir geht es darum, dass sie ihren Standpunkt in einem breiter gefächerten Umfeld finden sollen. Stellt euch und eure Arbeit in größere Zusammenhänge!«

»My assumption was that young people need help (also of a financial kind) in order to broaden their horizon. It's important for me that they should find out where they stand in a more diverse environment. Place yourselves and your work in a wider context!«

Mia Seeger, 1986

Ereignissen in den 20er und frühen 30er Jahren und als Verantwortliche in den 50er und 60er Jahren. Obwohl sie keine Designerin war, sondern eher in der Design-Vermittlung im Rat für Formgebung und im Deutschen Werkbund wirkte, wurde sie vielfach zu Triennalen in Mailand als Kommissarin der Bundesrepublik Deutschland für die deutsche Abteilung und nationalen und internationalen Designwettbewerben als Preisrichterin eingeladen. Ihr Urteil galt. Ihr gelang es als Frau, sich in der festgefügten Männerwelt des Design zu behaupten, geschätzt und verehrt zu werden.

Sie wirkte politisch im besten Sinne des Wortes: sie trug dazu bei, das deutsche Design nach der unsäglichen Naziherrschaft und dem verheerenden Zweiten Weltkrieg im Ausland wieder bekannt zu machen und neue Wege der Verständigung zu gehen. Sie war eine faszinierende Persönlichkeit. Als Mia Seeger alt war, nannte man sie »die große alte Dame des Deutschen Werkbundes« und »die große alte Dame des Design«. Sie war eine

Mia Seeger was born on 9 May 1903 in Bad Cannstatt, which today is part of Stuttgart. She died aged 88 on 14 May 1991, also in Stuttgart. She devoted her life to what we would call design in the widest sense of the word. She had the best teachers, including Mies van der Rohe and Walter Gropius. She witnessed the most important events of the 1920s and early 1930s as a student of design, and those of the 1950s and 1960s as someone responsible for promoting design. Although she was not a designer, but worked more as a communicator of design in the »Rat für Formgebung« and the »Deutscher Werkbund«, she was frequently invited to judge world exhibitions as well as national and international design competitions. Her verdicts had authority. And as a woman, she managed to assert herself and to be esteemed and respected in the entrenched, male-dominated world of design.

She acted politically in the best sense of the word: she helped make German design known again in the world following the unspeakable Nazi regime and the devastation of the second world war, and help new international understanding to arise. She was a fascinating personality. When she grew old, she was called the »grand old lady of the Deutscher Werbund« and the »grand old lady of design«. She was a warm-hearted, endearing women. To be recognized by her was an honour. For the ruling body of the Mia Seeger Foundation, she is a role model.

Looking at the products and projects commended by the Mia Seeger Foundation judges over the past 25 years, one is struck by a certain preference for the social and medical sphere. The principle of only awarding a prize to something that »benefits more than the individual« is always at the forefront. When Mia Seeger was herself still a judge, she decided that the prize should initially be offered to university graduates from Baden-Württemberg. With the design exchange and German reunification, its scope was extended to all universities of design in Germany. The spectrum of studies submitted is a wide one. Many graduates have concerned themselves with emergency aid, both water- and warmherzige, liebenswerte Frau, deren Anerkennung ehrte. Ihrem Vorbild fühlen sich die Gremien der Mia Seeger Stiftung verpflichtet.

Betrachtet man die durch die Jurys der Mia Seeger Stiftung ausgezeichneten Produkte oder Vorhaben nach 25 Jahren, so fällt eine gewisse Präferenz für den sozialen und medizinischen Bereich auf. Der Grundsatz, nur etwas auszuzeichnen, »was mehr als einem nützt«, steht dabei immer im Vordergrund. Als Mia Seeger noch selbst als Jurorin mitwirkte, bestimmte sie, den Preis zunächst für baden-württembergische Hochschulabsolventen auszuschreiben. Mit der Designbörse – und der Wiedervereinigung – kam die Ausweitung auf alle Designhochschulen in der Republik. Breit ist das Spektrum der eingereichten Arbeiten. Viele Absolventen befassen sich mit Katastrophenhilfe zu Wasser und aus der Luft, andere engagieren sich im ökologischen Bereich. Hightech steht Lowtech gegenüber, in jedem Fall ist der Einsatz an Ideen begeisternd, bewundernswert und preiswürdig. Inzwischen verzichtet die Stiftung auf formale Abschlüsse der Bewerber. Allein die Qualität der Arbeit zählt.

Immer wieder versichern frühere Preisträger und Stipendiaten, wie wichtig die Auszeichnung der Mia Seeger Stiftung für ihren Berufsstart war. Einige von ihnen sind aus dieser Verbundenheit heraus Zustifter geworden. Vor kurzem hat die Stiftung speziell für die Alumni das Minimum des einmaligen Beitrags auf 1.000 Euro gesenkt und ihre Bemühungen um Zuwendungen aus der Industrie verstärkt. So hofft sie, weiterhin den Nachwuchs auf seinem Weg zu verantwortlicher Gestaltung nicht nur mit Anerkennung loben, sondern auch mit ordentlichem Preisgeld belohnen zu können. Und sie vertraut darauf, dass ein erweiterter und verjüngter Beirat das Erbe Mia Seegers ideenreich und in ihrem Geiste fortführt.

Prof. Karin Kirsch
Beiratsvorsitzende der Mia Seeger Stiftung, Stuttgart

air-borne, while others have dealt with ecological issues. High-tech meets low-tech, but at all events, the ideas presented are inspiring, admirable, and worthy of commendation. The foundation now no longer requires entrants to have formal qualifications. All that counts is the quality of the work.

Again and again, former prize-winners and scholarship holders underscore how important a commendation by the Mia Seeger Foundation was for their career start. In some cases, this has also resulted in donations to the foundation. Especially for the alumni, the foundation recently cut its minimum one-off contribution to 1,000 euros. It has also increased its fund-raining efforts among companies. In this way, it hopes to be able to reward young designers progressing towards responsible design not only with warm words but also with adequate prize money. And it puts its trust in an extended, rejuvenated advisory committee, which will continue Mia Seeger's legacy imaginatively and in her spirit.

Prof. Karin Kirsch
Chair of the Mia Seeger Stiftung advisory board, Stuttgart

Mia Seeger Preis 2011

Jährlicher Wettbewerb der Mia Seeger Stiftung für junge Designerinnen und Designer

Mia Seeger Prize 2011

Mia Seeger Foundation annual competition for young designers

»Das war eigentlich meine Idee – beim Nachdenken über den Preis: in die »Lücken« gehen, die noch nicht so beachtet worden sind.«

»When I had the idea of a prize, what really concerned me was to recognize work done in areas that have tended to be neglected up to now.«

Mia Seeger, 1987

Wer auf der Suche nach speziellen Informationen zur Stiftung und ihren Aktivitäten oder beim Wunsch, die Stiftung mit einer Spende oder Zustiftung zu unterstützen, den Kontakt zur Geschäftsführung braucht, wendet sich an die Geschäftsstelle:

Anyone wishing to contact the management of the foundation, either because they are interested in learning more about the foundation and its activities, or wish to support the foundation with a donation or endowment, should contact the foundation office:

Mia Seeger Stiftung
c/o Design Center Stuttgart
im Haus der Wirtschaft
Willi-Bleicher-Straße 19
D-70174 Stuttgart
T +49 711 123 2681
F +49 711 123 2771
E-Mail: design@rps.bwl.de
www.mia-seeger.de

Zur Teilnahme eingeladen waren Studenten und Absolventen an deutschen Hochschulen aus gestalterischen Studiengängen mit Studien- und Abschlussarbeiten, die in den Jahren 2009 bis 2011entstanden sind. Bei der Bewertung der eingereichten Entwürfe ist neben den üblichen Designkriterien der soziale Nutzen entscheidend.

Ausgelobt waren 10.000 Euro

Jury
Prof. Karin Kirsch
Beiratsvorsitzende der Mia Seeger Stiftung, Stuttgart

Margarete Wies
Mercedes-Benz Design,
Leiterin Mercedes-Benz Style & Design Strategy,
Sindelfingen

Prof. Ulrich Hirsch
Design & Identität, Brügge,
Schleswig-Holstein

Peter Keilbach
1991 Preisträger und Stipendiat der Mia Seeger Stiftung,
keilbach design and products,
Dörzbach

Leo Lübke
Geschäftsführender Gesellschafter COR und interlübke, Rheda-Wiedenbrück

Students and graduates of design-related courses at German universities were invited to enter course work and dissertations produced between 2009 and 2011. Apart from the usual design criteria, the decisive benchmark when assessing the concepts submitted is their social utility.

Total prize money: 10,000 euros.

Judges panel
Prof. Karin Kirsch, chair of the Mia Seeger Stiftung advisory board, Stuttgart

Margarete Wies
Mercedes-Benz Design, director of Mercedes-Benz style and design strategy, Sindelfingen

Prof. Ulrich Hirsch
Design & Identität, Brügge

Peter Keilbach
Mia Seeger Prize and scholarship 1991,
keilbach design and products, Dörzbach

Leo Lübke
managing partner of the furniture manufacturers COR and interlübke, Rheda-Wiedenbrück

Sigecco
Temporäres Fixieren auf Großbaustellen

Temporary anchorage point for construction sites

Mia Seeger Preis 2011
1. Preis

Mia Seeger Prize 2011
1st prize

Jury
Judges panel

Auf Baustellen, großen zumal, kreuzen sich Feines und Grobes, Intelligibles und Hemdsärmeliges. In »Sigecco« umschließen sie einander konzentrisch: in der Mitte Andruck (von vorn) und Festhalten (per Nano auf der Rückseite), dann ein Ring für Information, auch für Warnung; und außen herum Schutz vor Stoß und Schmutz.

Zusammen mit Spann- und Lasthebegurten, Gurtmanschetten, Klemmleisten, Leuchten, Schildern und Montageschablonen löst der Entwurf des Anschlagpunkts gravierende Probleme des Baubetriebs: stiftet Ordnung, rationalisiert Abläufe, sorgt für Sicherheit.

On construction sites, especially large ones, there is a hotchpotch of the rough and the refined, of blue-collar and white-collar work. In Sigecco, they embrace each other in concentric circles: at the centre, pressure is applied (from the front) and the point sticks (thanks to nanotechnology on the rear surface). This is followed by a ring for information and also for warning, and an outer ring providing protection against impact and dirt.

Together with lashing and lifting belts, belt collars, terminal strips, lamps, signs and assembly templates, this study of an anchorage point solves some difficult problems encountered on building sites: it creates order, rationalizes processes, and provides safety.

For construction workers, a hook for hanging things on would be very convenient. If there were at least a palm-sized disc that stayed put anywhere it was fastened on the carcase of a building, from which all the cables, pipes, barriers, signs, lamps, tools could be hung that are needed temporarily during building work, with belts to keep them in place. A building expert would call such a thing an anchorage point.

Researchers and engineers are now close to perfecting adhesive films that can imitate the adhesive force of geckos' toe pads. Equipped with such films, Sigecco brings the dream of an anchorage point that will stick anywhere close to realization.

Entwerfer
Developer
Peer Büscher
Bahnhofstraße 14
D-64372 Ober-Ramstadt
M +49 1 78 7 46 47 56
peer.buescher@design-buerozwo.de

Hochschule
Industriedesign,
Diplom 2010
Hochschule Darmstadt
University
University of Applied Sciences Darmstadt

Betreuung
Support
Prof. Tom Philipps

Ein »Lufthaken« käme den Bauleuten sehr zupass. Wenn es wenigstens eine handtellergroße Scheibe gäbe, die überall an Wand und Decke im Rohbau hielte, um daran all die Kabel, Leitungen, Absperrungen, Schilder, Lampen, Gerätschaften, die der Baubetrieb vorübergehend braucht, zu hängen, dazu Gurte als Haltehilfen. »Anschlagpunkt« nennt es der Mann vom Fach.

Forscher und Techniker sind jetzt mit Klebefolien nah dran, die Haftkraft an den Füßen des Geckos nachzuahmen. Damit ausgerüstet, hebt »Sigecco« den Traum vom Halt aus der Luft nahezu ins Machbare.

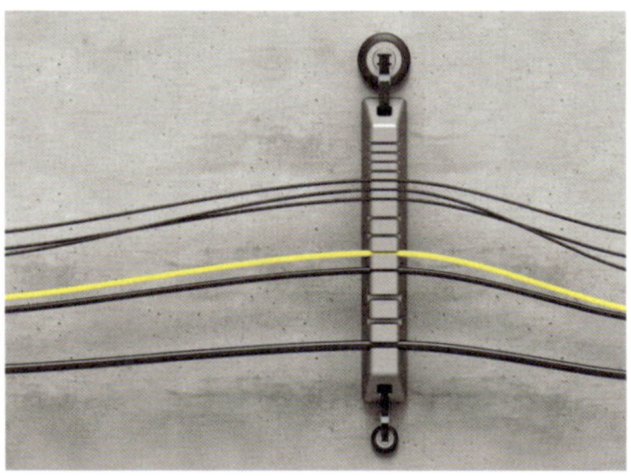

Jury
Judges panel

Wolt
Windkraft im Weinberg

Wind power in the vineyard

Mia Seeger Preis 2011
1. Preis

Mia Seeger Prize 2011
1st prize

Gestalterisch zurückgenommen, integriert sich »Wolt« in Dimension und Betrieb perfekt in vorhandene Weinberganlagen und stört weder den Weinbau noch das Landschaftsbild. Durchweg wurde beim Entwurf auf einfache, weitgehend selbst erprobte Technik und kostengünstige Herstellung geachtet, und darauf, dass die Anlage mit geringem Aufwand zu installieren, überwachen, warten, reparieren und recyclen ist. Mit findiger Sorgfalt, bescheiden, aber unmissverständlich ist bezeichnet, wie Energiewende auch aussehen kann.

»Wolt« harmonizes perfectly with existing vineyards, both in its size and its operation. It does not interfere either with the vintner or with the scenery. At all times, the study pays attention to simple technology that the designer has himself tried out in most cases, as well as to cost-effective manufacture. It also ensures that the system can be installed, monitored, maintained, repaired and recycled at little cost. Ingenious and modest, but with an unmistakable message that a move to new energies can also look like this.

Windpower from the vineyard? Using hundreds of elements based on windbelt technology, it is possible. A narrow belt to which two magnets have been attached is stretched over a long bow. When the wind causes the band to vibrate, the magnets plunge into coils and induce alternating current. After passing through an inverter, this current is fed into the windbelt circuit. These wind power elements can easily be attached to vineyard posts using cable ties. The electric cables are directed along the wires used to support the vines.

Entwerfer
Developer
Marko Müller
Hohlweg 8
D-54316 Hockweiler
M +49 1 79 7 56 89 60
muellermarko@gmx.de

Hochschule
Industriedesign,
Diplom 2010
Hochschule Darmstadt
University
University of Applied
Sciences Darmstadt

Betreuung
Support
Prof. Tino Melzer

Windkraft aus dem Weinberg? Mit Hunderten von Elementen, die auf der »Windbelt«-Technologie gründen, ist es möglich. Über einen langen Bügel ist ein schmales Band gespannt; daran sind zwei Magnete befestigt. Wenn der Wind das Band zum Flattern bringt, tauchen die Magnete in Spulen ein und induzieren Wechselstrom, der gleichgerichtet in die Verschaltung der Windbelts eingespeist wird. Die Windenergie-Elemente lassen sich einfach mit Kabelbindern an den Weinbergpfosten befestigen, und die Verkabelung wird an den Spanndrähten entlanggeführt.

Bambino
Ein pädiatrisches Krankenbett
A pediatric hospital bed

Mia Seeger Preis 2011
Anerkennung
Highly commended

Entwerfer
Developer
Daniela Colaianni
Voigtländerstraße 1
D-90489 Nürnberg
M +49 1 76 24 81 44 66
mail@danielacolaianni.de

Hochschule
Integriertes Produkt-design, Diplom 2011
Hochschule Coburg
University
University of Applied Sciences Coburg

Betreuung
Support
Prof. Peter Raab
in Kooperation mit
Siemens AG

Jury
Judges panel

Materialien, Konfiguration und technische Ausstattungen sind so bemessen, dass sie im Krankenhausbetrieb leisten, was er verlangt, auch im Blick auf die Kosten. Dem behandelnden Personal werden Arbeitsaufwand und -ablauf erleichtert. Vor allem aber wurde Wert darauf gelegt, dass Eltern bei ihrem kleinen Patienten sein können, lang am Tag und über Nacht. Die Formensprache der Möbel und das Farbkonzept Weiß-Grün-Braun vermitteln eine frische Atmosphäre heiterer Zuversicht: »... wohlan, mein Kind, gesunde!«

Materials, configuration and technical equipment have been gauged to provide what is demanded in everyday hospital routine, also with respect to costs. The workload and workflow of hospital personnel have been eased. The main concern, however, was that parents should be able to be with their tiny patients, for a long time during the day and also at night. The furniture's formal expression and the white-green-brown colour scheme create a fresh atmosphere of cheerful confidence: »... chin up, child, and get better soon«.

Ein Krankenhaus ist keine Reparaturanstalt, erst recht nicht für Kinder. Das fängt am Bett an. Besser, es kann in Länge, Höhe und Liegegeometrie auf ihre Körpergröße, Behandlung und Bedürfnisse ver- und eingestellt werden, und eine textile, durchscheinend verwobene Seitenbespannung bietet Sicherheit. Es passt sich an der Kopfseite ein in ein wandmontiertes Kastenmöbel, das verschiedene Ablagen vorsieht, alle Medienanschlüsse enthält, dazu Stauraum für medizinisches Material und persönliche Sachen, außerdem eine ausklappbare Liege, auf der ein Elternteil sitzen, ruhen, auch beim Kind übernachten kann. Wenn das Kinderbett nicht belegt ist, verschwindet es zusammengeklappt fast ganz in seiner Nische.

A hospital is not a repair shop, certainly not for children. This starts with hospital beds. It is better to have beds whose length, height and horizontal geometry can be adjusted and adapted to children's body size, treatment and needs. A semi-transparent textile side netting offers safety. At the head end, it fits into a wall-mounted piece of cabinet furniture that has various shelves and connections for media of all kinds, including stowage space for medical materials and personal effects. It also has a fold-out daybed on which a parent can sit, relax or spend the night with the child. If the child's bed is not in use, it is folded up and practically disappears into its niche.

First Aid
Hilfe für den Helfer
Help for helpers

Mia Seeger Preis 2011
Anerkennung
Highly commended

Entwerfer
Developer
Carine Collé
38, Rue E. Servais
L-4989 Sanem
Luxemburg
M +352 15 77 5 11 31 61
ca.colle@gmail.com

Hochschule
Industrial Design,
Bachelor of Arts 2011
Hochschule Pforzheim
University
University of Applied
Sciences Pforzheim

Betreuung
Support
Prof. Jürgen Goos und
Prof. Dipl.-Ing. Ralph
Schieschke

Jury
Judges panel

Die Anpassung an den Körperbau machen bei Weste und Gürtel eine komplexe Struktur elastischer und zugfester Bänder, stützender und polsternder Stoffe, belastbarer Armierungen, verstellbarer Schlaufen und klickender Verschlüsse notwendig. Moderne Materialien und vielfach verfügbare Accessoires erleichtern die Aufgabe. Indem die Entwerferin dem Weg der Retter in allen Eventualitäten und Beschwernissen nachspürte, hat sie ihnen eine gesundheitlich zuträglichere Lösung vorgeschlagen – eine, die sich auch auf gängige Tragbahren übertragen lässt.

Adjusting to body size means that a complex structure of elastic and tensile bands, supporting and upholstering textiles, hard-wearing reinforcements, adjustable straps and click fasteners are needed in the vest and belt. Modern materials and readily available accessories make this task easier. The designer has thought of every eventuality and difficulty on the paramedics' path, and in this way has come up with a solution that is better for their physical health, and that can be applied to conventional stretchers.

Doppelt schwer hat's der Rettungsdienst, wo er den Kranken zu Fuß tragen muss. Da hilft es, wenn der vordere Träger die Last am Rücken in eine schienenbewehrte Weste einhängen und gelegentlich, z.B. an Türen, einen Handgriff loslassen kann. Der hintere Träger aber erfährt Entlastung nur durch einen Bauchgurt. Er muss mit beiden Händen an den Griffen bleiben, den Getragenen überwachen und den Transport dirigieren. Um besonders enge Situationen, in Treppenhäusern z.B., meistern zu können, sind die Griffgestelle nach Sattelschlepperart mit der Trage verbunden.

The paramedics' task is twice as hard if they have to carry a patient on foot. It is a help if the carrier at the front can bear the weight on his back using a vest that is reinforced with rails, and if he can occasionally let go of a handle when he has to open a door, for example. For the carrier at the back, a waist belt takes some of the weight. He has to keep both hands on the handles, keep a watch on the patient, and direct the transport. To be able to master especially tight corners – for example on stairs – the handle frame has been connected to the stretcher like a trailer to an articulated lorry.

De Care
Decubitus Healthcare

Mia Seeger Preis 2011
Anerkennung
Highly commended

Entwerfer
Developer
Cristian Costoiu
Dechant-Meyer-Straße 26
D-49377 Vechta
M +49 1 77 9 75 78 66
cristiancostoiu@msn.com

Hochschule
Industriedesign, Bachelor of Arts 2011
Muthesius Kunsthochschule Kiel
University
Muthesius Academy of Fine Arts and Design Kiel

Betreuung
Support
Prof. Thomas Feichtner und
Prof. Dr. Norbert Schmitz
in Kooperation mit
Porsche Design Studio

Jury
Judges panel

Erstaunlich, welche Menge an Steuerelektronik, Verstellmechanik und Servotechnik in ein derart schlankes Gestell und in so flache Polsterschalen hineinpasst. Für Bedienelemente wie Interface, Schiebebügel oder Stellknöpfe sind flächenbündige Lösungen gefunden. Nichts beeinträchtigt die klaren Konturen des Möbels, das sich in Linienführung, Farbgebung und Materialwahl jederzeit in ein modernes wohnliches Umfeld einfügt. Alles ist auf hochwertige Anmutung gerichtet. Nichts stigmatisiert den Nutzer.

Amazing: the quantity of control electronics, adjusting mechanisms and servo technology that will fit into such a slim frame and such flat upholstered cushions. The operating elements such as interface, push handle and adjusting knobs have been flush mounted. Nothing gets in the way of the furniture's clear contours. Its lines, colour scheme and choice of materials mean it will harmonize with any modern living environment. Everything is directed towards an air of high quality. There is nothing to stigmatize the user.

Wund vom Liegen. Da verlangt es den Dekubituskranken doch sehr nach einer Matte, die ihn an befallenen oder gefährdeten Stellen entlastet, den Druck umverteilt und sonst gesundes Gewebe rhythmisch stimuliert. Die EAP-Forschung in der Schweiz macht ihm berechtigte Hoffnung auf Linderung mit einer Matrix aus kleinen, einzeln ansteuerbaren elektroaktiven Polymerkissen. Sie ist als Auflage für einen Pflegesessel gedacht, der im übrigen so konzipiert ist, dass der Patient in alle Lagen zwischen Liegen, Sitzen und Stehen gebracht und nach Bedarf gefahren werden kann.

Bed sore: the patient suffering from decubitus will long for a mat that takes the weight off the parts that are sore or in danger of becoming so. Such a mat should redistribute pressure and rhythmically stimulate otherwise healthy tissue. EAP research in Switzerland holds out a promising prospect of relief, with a matrix of tiny polymer cushions, each one of them individually controllable. It is conceived as a cushioned cover for a nursing chair. This chair is designed so that the patient can be placed in any position between supine, seated and standing, as required.

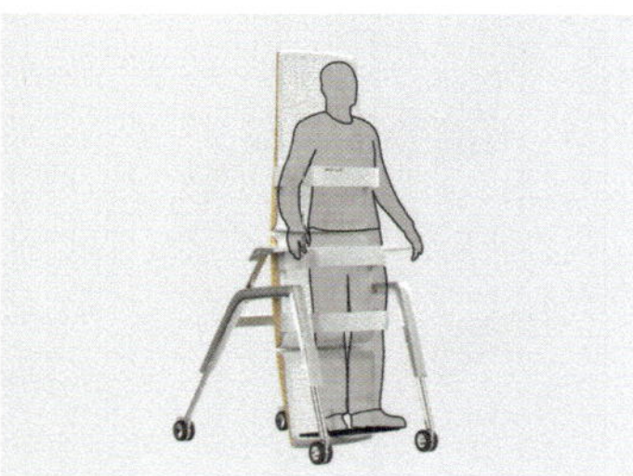

Emergency Airdrop
Hilfsgüterabwurf
Airdrop for relief goods

Mia Seeger Preis 2011
Anerkennung
Highly commended

Entwerfer
Developer
Adrienne Finzsch
Spessartring 2
D-64287 Darmstadt
M +49 1 60 4 09 50 89
afinzsch@aol.com

Hochschule
Industriedesign,
Vordiplom 2010
Hochschule Darmstadt
University
University of Applied
Sciences Darmstadt

Betreuung
Support
Prof. Tom Philipps

Jury
Judges panel

Beste Bionik, verbunden mit Transformationstechnik und Pop-up-Mechanik. Schritt für Schritt ist die Grundidee nach den Gegebenheiten des Hilfseinsatzes entfaltet und ausgestaltet: eine Packungsgeometrie, die den Frachtraum besser nutzt; Verpackungsmaterial aus Altpapier, das durch Wachsbeschichtung wetterbeständig wird; Aufschrift und Symbolik, die den Inhalt kennzeichnen und die Verteilung erleichtern.

Am wichtigsten aber ist, dass der passive Schraubenflug die bisher hohen Verluste beim Abwurf von Hilfsgütern drastisch vermindern kann.

Excellent bionics, combined with transformation technology and a pop-up mechanism. Step by step, the basic idea has been developed, taking the conditions of disaster relief into account. The geometry of the packaging allows the hold to be better used, the packaging is made of recycled paper, made weatherproof by coating it with wax, and the labelling and symbols clearly describe the contents and make distribution easier.

Most importantly, however, the passive propeller flight can dramatically reduce losses when aid is dropped, which have been high up to now.

Eine Verpackung mit Flügeln – wie bei den Samen von Ahorn oder Linde. Dann stürzt die wertvolle Fracht, wenn sie vom Flugzeug aus über Katastrophengebieten abgeworfen wird, nicht wie ein Stein vom Himmel, sondern trudelt im stabilen Rotationsflug zu Boden. Zum Transport wird der eigentliche Container in die geflügelte Umverpackung eingehängt. Wie Hasenohren stellen sich da die Flügel auf. Beim Abwurf breiten sie sich wieder aus. Zwischen Containerboden und Umverpackungsboden stellt sich wieder der alte Abstand her. Das dämpft den Aufprall bei der Landung.

A packaging with wings – like the fruit of a sycamore or lime tree. When the valuable aid packages are dropped from an aeroplane over disaster areas, they do not plummet like a stone, but spin to the ground in a stable flight pattern. To transport the goods, the actual container is placed inside the winged outer packaging. The wings stand upright like the ears of a hare. When thrown from the aeroplane, the wings spread again. A gap is created between the floor of the container and the floor of the outer packaging. This cushions the impact of landing.

Elliot
Rettungsfahrzeug mit Ersthelfer-station für Katastrophengebiete
Ambulance with first-aid station for disaster areas

Mia Seeger Preis 2011 Anerkennung
Highly commended

Entwerfer
Developer
Stephanie Knödler
Adelmannweg 6
D-73579 Schechingen
M +49 1 76 22 50 93 95
stephanie.knoedler@gmx.de

Hochschule
Produktgestaltung,
Bachelor of Arts 2011
Hochschule für Gestaltung Schwäbisch Gmünd
University
University of Applied Sciences Schwäbisch Gmünd

Betreuung
Support
Prof. Dr. Jürgen Held
und
Prof. Peter Stebbing

Jury
Judges panel

Die Aufgabenstellung leitet sich ab aus der Einsicht, dass Katastrophenhilfe wesentlich ein Problem der Infrastruktur ist. Darum widmet sich der Entwurf einem Fahrzeug, das die Hilfsmittel der medizinischen Versorgung mitbringt und dann weiter als logistisches Vehikel zur Verfügung steht. Damit baut er an der Gesamtstruktur, sich in diese einfügend, zugleich mit. Mit seiner kompakten Bauweise und vielen detaillierten Ausstattungsmerkmalen ist das Fahrzeug auf einen flexiblen Einsatz unter unwirtlichen Umständen hin konzipiert.

The task is guided by the insight that relief work is mainly a problem of infrastructure. This is why the study considers a vehicle that transports the medical aid and is then available for logistical purposes. This influences the vehicle's overall structure, and is integrated into it. With its compact design and many features and details, the vehicle is designed for flexible operation in inhospitable conditions.

Schlimm sind großflächige Katastrophen auch deshalb, weil die Wege zu den Opfern weit und unwegsam sind. Es geht also nicht ohne ein geländegängiges Rettungsfahrzeug. Es fährt nahe an die Schwerverletzten heran. Die Helfer errichten mit dem mitgebrachten Container und dem darin mitgeführten medizinischen Material außerhalb des Fahrzeugs eine Ersthelfer-Station, die dann noch mit Schutzdach und Trinkwasseraufbereitung versehen werden kann. Danach dann dient das Fahrzeug dem Transport von geborgenen Verletzten zur Station und, sobald sie dort provisorisch versorgt sind, weiter zum Not-Hospital.

One reason why large-scale disasters are so terrible is that victims are remote and the roads almost impassable. Any ambulance therefore has to be suitable for off-road driving. This ambulance drives up to the badly injured victims. Using the container they have transported and the medical materials inside, the helpers set up a first-aid station, which can also be provided with a protective roof and a water treating facility. The vehicle can then be used to transport rescued victims to the station and, as soon as they have received initial treatment, on to the emergency hospital.

Impressum

Herausgeber:
Mia Seeger Stiftung
Redaktion:
Dr. Wolfgang Berger
Übersetzung:
Philip Mann, Marbach
Grafikdesign:
stapelberg & fritz, Stuttgart
Roman Heinrich
Ausstellungsgestaltung:
Sandy Tsiris, sitibi
Kommunikation GmbH, Stuttgart
Fotos:
Preisträger und Ausgezeichnete

Stuttgart 2011

Publishing details

Published by:
Mia Seeger Stiftung
Editor:
Dr. Wolfgang Berger
Translation:
Philip Mann, Marbach
Graphic design:
stapelberg & fritz, Stuttgart
Roman Heinrich
Exhibition design:
Sandy Tsiris, sitibi
Kommunikation GmbH, Stuttgart
Photographs:
prize-winners and commended entrants

APPENDIX

Adressen
Hersteller, Designer und Vertriebe

Addresses
Manufacturers, designers distributors

Namenregister

Index of names

Fotonachweis

Photographs (acknowledgements)

242
243

Adressen
Addresses

Hersteller, Designer und Vertriebe

Manufacturers, designers and distributors

A

Ahape Design
Lise-Meitner-Straße 21
D-72202 Nagold
T +49 74 52 8 37 00 41
F +49 74 52 8 37 00 44
kontakt@ahape.de
www.ahape.de
Seite
Page 38

Ambulanz Mobile GmbH & Co. KG
Glinder Straße 1
D-39218 Schönebeck
T +49 39 28 48 63 30
F +49 39 28 48 63 32
info@ambulanzmobile.eu
www.ambulanzmobile.eu
Seite
Page 212

Atelier Brückner GmbH
Krefelder Straße 32
D-70376 Stuttgart
T +49 7 11 5 00 07 71 26
F +49 7 11 50 00 77 22
presse@atelier-brueckner.de
www.atelier-brueckner.de
Seite
Page 196 ff, 200

Aturon GmbH
Seestraße 14
CH-9323 Steinach
T +41 71 3 51 47 61
F +41 71 3 51 47 62
mh@aturon.ch
www.aturon.ch
www.moonlet.com
Seite
Page 144

Werksdesign
In-house design
Marcus Höglmeier
Seite
Page 144

Auerhahn Bestecke GmbH
Im oberen Teil 9
D-72213 Altensteig
T +49 74 53 946 80
F +49 74 53 94 68 90
info@auerhahn-bestecke.de
www.auerhahn-bestecke.de
Seite
Page 90

B

Jacob Bachmayer
Nordbahnhofstraße 191
D-70191 Stuttgart
T +49 7 11 8 82 04 98
F +49 7 11 81 04 08 31
jacob@renschler.de
www.ingenieur-design.com
Seite
Page 50, 52

Bär GmbH
Pleidelsheimer Straße 15
D-74321 Bietigheim-Bissingen
T +49 71 42 95 66 85
F +49 71 42 95 66 66
christof.baer@baer-schuhe.de
www.baer-schuhe.de
Seite
Page 162

Werksdesign
In-house design
Michael Bach
Seite
Page 162

Bar Vinya Film
Jan Reiff
Auf der Scherersmatt 4
D-77797 Ohlsbach
info@filmreiff.de
www.filmreiff.com
Seite
Page 148

Becker GmbH
Schmelzofenvorstadt 33/4
D-89520 Heidenheim
T +49 73 21 3 57 30
F +49 73 21 35 73 35
t.becker@agentur-becker.de
www.agentur-becker.de
Seite
Page 130

Bene AG
Schwarzwiesenstraße 3
A-3340 Waidhofen/Ybbs
T +43 74 42 5 00 34 97
sabine.roseneder@bene.com
www.bene.com
Seite
Page 96

Bene Stuttgart
Breitwiesenstraße 19
D-70565 Stuttgart
T +49 7 11 3 41 81 80
sabine.roseneder@bene.com
www.bene.com
Seite
Page 96

Bosch Thermotechnik GmbH
Junkersstraße 20-24
D-73249 Wernau
T +49 71 53 3 06 27 76
F +49 8 11 5 16 13 55
christian.buettner2@de.bosch.com
www.bosch-thermotechnik.de
Seite
Page 186 ff

Werksdesign
In-house design
Fusion-x Community Inhouse
Seite
Page 186 ff

Bosch Thermotechnik GmbH
Junkers Deutschland
Postfach 1309
D-73243 Wernau
T +49 18 03 3 37 33
F +49 18 03 33 73 32
junkers.infodienst@de.bosch.com
www.junkers.com
Seite
Page 186 ff

Ronan und Erwan Bouroullec
23, Rue du Buisson St. Louis
F-75010 Paris
T +33 1 42 00 40 33
F +33 1 42 00 52 11
info@bouroullec.com
www.bouroullec.com
Seite
Page 70

Burkhardt Leitner constructiv GmbH & Co. KG
Breitwiesenstraße 17
D-70565 Stuttgart
T +49 7 11 2 55 88 13
F +49 7 11 2 55 88 11
olp@burkhardtleitner.de
www.burkhardtleitner.de
Seite
Page 204

Werksdesign
In-house design
Team Burkhardt Leitner constructiv
Seite
Page 204

Peer Büscher
Bahnhofstraße 14
D-64372 Ober-Ramstadt
M +49 1 78 7 46 47 56
peer.buescher@design-buerozwo.de
Seite
Page 232

Busch-Jaeger Elektro GmbH
Freisenberg 2
D-58513 Lüdenscheid
T +49 23 51 9 56 17 77
bernhard.heitz@de.abb.com
www.busch-jaeger.de
Seite
Page 206

C

Cap. GmbH
Tonstraße 25
D-32609 Hüllhorst
T +49 57 44 5 08 20
F +49 57 44 50 82 10
info@cap-direct.de
www.cap-direct.de
Seite
Page 92

Werksdesign
In-house design
Freimut Stehling
Seite
Page 92

Franco Clivio
Rietstraße 11
CH-8703 Erlenbach
T +41 44 9 12 06 60
fclivio@bluewin.ch
Seite
Page 184

Daniela Colaianni
Voigtländerstraße 1
D-90489 Nürnberg
M +49 1 76 24 81 44 66
mail@danielacolaianni.de
Seite
Page 236

Carine Collé
38, Rue E. Servais
L-4989 Sanem (Luxemburg)
M +352 15 77 5 11 31 61
ca.colle@gmail.com
Seite
Page 237

Cristian Costoiu
Dechant-Meyer-Straße 26
D-49377 Vechta
M +49 1 77 9 75 78 66
cristiancostoiu@msn.com
Seite
Page 238

D

Daimler AG
HPC X 800
D-71059 Sindelfingen
T +49 70 31 9 04 42 34
F +49 70 31 90 36 99
hansharald.hanson@daimler.com
www.daimler.com
Seite
Page 216 ff, 220, 222,224

Designaffairs GmbH
Rosenheimer Straße 145b
D-81671 München
T +49 89 4 42 32 91 17
F +49 89 4 42 32 92 00
stefan.hillenmayer@designaffairs.com
www.designaffairs.com
Seite
Page 110

Ding 3000
Ratswiese 18
D-30453 Hannover
T +49 5 11 3 53 93 76
F +49 5 11 3 53 93 76
info@ding3000.com
www.ding3000.com
Seite
Page 112

Duravit AG
Werderstraße 36
D-78132 Hornberg
T +49 78 33 7 04 61
F +49 78 33 70 74 65
info@duravit.com
www.duravit.com
Seite
Page 66 ff

Dürr Dental AG
Hopfigheimer Straße 17
D-74321 Bietigheim-Bissingen
T +49 71 42 70 50
F +49 71 42 70 54 41
heiner.k@duerr.de
www.duerr.de
Seite
Page 44 ff

Dyson GmbH
Lichtstraße 43b
D-50825 Köln
T +49 2 21 50 60 00
F +49 2 21 50 60 01 90
infoline@dyson.de
www.dyson.de
Seite
Page 114

Dyson Ltd.
Tetbury Hill
Malmesbury,
Wiltshire SN 16 ORP UK
T +44 16 66 82 7 20 00
www.dyson.co.uk
Seite
Page 114

Werksdesign
In-house design
James Dyson
Seite
Page 114

E

Emamidesign
Arman Emami
Reinhardtstraße 6
D-10117 Berlin
T +49 30 47 98 76 54
F +49 30 47 98 76 55
info@emamidesign.de
www.emamidesign.de
Seite
Page 104, 138

EOOS Design GmbH
Zelinkagasse 2/6
A-1010 Wien
T +43 1 4 05 39 87
F +43 1 4 05 39 87 80
design@eoos.com
www.eoos.com
Seite
Page 66 ff

Erco GmbH
Brockhauser Weg 80-82
D-58507 Lüdenscheid
T +49 23 51 55 10
F +49 23 51 55 13 00
info@erco.com
www.erco.com
Seite
Page 128

Werksdesign
In-house design
Erco Design Team
Seite
Page 128

Esselte Leitz GmbH & Co. KG
Siemensstraße 64
D-70469 Stuttgart
T +49 7 11 8 10 35 05
F +49 7 11 8 10 38 26
pdingethal@esselte.de
www.leitz.de
Seite
Page 146

Eva Solo A/S
Måløv Teknikerby 18-20
DK-2760 Måløv
T +45 73 20 60
F +45 70 74 11
sif@evasolo.com
www.evasolo.com
Seite
Page 80, 82, 108

F

Fackelmann GmbH & Co. KG
Sebastian-Fackelmann-Straße 6
D-91217 Hersbruck
T +49 91 51 81 11 24
F +49 91 51 8 11 40
martina.miller@fackelmann.de
www.fackelmann.de
Seite
Page 84

Werksdesign
In-house design
Fabian Graf von Reichenbach
T +49 91 51 81 11 30
fabian.reichenbach@fackelmann.de
Seite
Page 84

Faro Scanner Production GmbH
Lingwiesenstraße 11/2
D-70825 Korntal-Münchingen
T +49 71 50 9 79 75 00
F +49 71 50 97 97 95 00
reinhard.becker@faroeurope.com
www.faro.com
Seite
Page 32

Festo AG & Co KG
Karoline von Häfen
Ruiter Straße 82
D-73734 Esslingen
T +49 7 11 34 70
www.festo.com
Seite
Page 42

Festo Didactic GmbH & Co. KG
Rechbergstraße 3
D-73770 Denkendorf
www.festo.com
Seite
Page 42

Adrienne Finzsch
Spessartring 2
D-64287 Darmstadt
M +49 1 60 4 09 50 89
afinzsch@aol.com
Seite
Page 239

Uwe Fischer
Industrial Design
Laubestraße 22
D-60594 Frankfurt
T +49 69 49 00 11
info@office-uwefischer.com
www.office-uwefischer.com
Seite
Page 126

Flux-Geräte GmbH
Talweg 12
D-75433 Maulbronn
T +49 70 43 10 10
F +49 70 43 10 11 11
info@flux-pumpen.de
www.flux-pumpen.de
Seite
Page 34

Formstudio Merklepark GbR
Rotebühlstraße 66
D-70178 Stuttgart
T +49 7 11 7 22 56 21 81
F +49 7 11 7 22 56 21 83
u.merkle@formstudio.de
www.formstudio.de
Seite
Page 44 ff

Freecom Technologies GmbH
Sachsendamm 3-4
D-10829 Berlin
T +49 30 6 11 29 90
F +49 30 6 11 65 88
info@freecom.de
www.freecom.de
Seite
Page 138

Fuchs Design
Alte Dorfstraße 26b
D-70599 Stuttgart
T +49 7 11 4 56 08 99
F +49 7 11 4 57 98 01
fuchsdesign@gmx.com
www.fuchsdesign.biz
Seite
Page 58

Fun Factory GmbH
Am Hohentorshafen 17/19
D-28197 Bremen
T +49 4 21 52 07 60
F +49 4 21 52 07 61 29
info@funfactory.de
www.funfactory.de
Seite
Page 165, 166

Werksdesign
In-house design
Nina Smerzinski
T +49 4 21 52 07 62 20
Seite
Page 165

Werksdesign
In-house design
Verena Michel
T +49 4 21 52 07 62 20
Seite
Page 166

G

Gerriets GmbH
Im Kirchenhürstle 5-7
D-79224 Umkirch
T +49 76 65 96 02 01
F +49 76 65 96 01 55
hgerriets@gerriets.com
www.gerriets.com
Seite
Page 28

Ghost Bikes GmbH
An der Tongrube 3
D-95652 Waldsassen
T +49 96 32 9 25 50
F +49 96 32 92 55 16
info@ghost-bikes.de
www.ghost-bikes.de
Seite
Page 176

Gira
Giersiepen GmbH & Co. KG
Dahlienstraße 12
D-42477 Radevormwald
T +49 21 95 60 20
F +49 21 95 60 21 14
info@gira.de
www.gira.de
Seite
Page 193

Gloria Haus- und Gartengeräte GmbH
Därmannsbusch 7
D-58456 Witten
T +49 23 02 70 00
F +49 23 02 7 00 46
a.werner@gloria-garten.com
www.gloriagarten.de
Seite
Page 110

Gloria Haus- und Gartengeräte GmbH
Vertriebs- und Marketingzentrale
Edisonallee 3
D-89231 Neu-Ulm
T +49 7 31 1 40 60 44
F +49 7 31 1 40 60 55
a.werner@gloria-garten.com
www.gloriagarten.de
Seite
Page 110

Grauwert
Büro für demografiefeste Produkte und Dienstleistungen
Mathias Knigge
Ruhrstraße 13
D-22761 Hamburg
T +49 40 79 69 37 57
F +49 40 79 69 37 58
mail@grauwert.info
www.grauwert.info
Seite
Page 158

Solveig Gubser
Grolmannstraße 22
D-10623 Berlin
M +49 1 76 82 13 97 51
solveiggubser@yahoo.de
Seite
Page 162

H

Haeberle GmbH & Co. KG
Breitwiesenstraße 13
D-70565 Stuttgart
T +49 7 11 7 83 14 37
F +49 7 11 7 83 14 77
info@haeberle-med.de
www.haeberle-med.de
Seite
Page 58

Hansgrohe AG
Auestraße 5-9
D-77761 Schiltach
T +49 78 36 5 10
F +49 78 36 51 13 00
info@hansgrohe.de
www.hansgrohe.de
Seite
Page 70

Hebumedical GmbH
Badstraße 8
D-78532 Tuttlingen
T +49 74 61 9 47 10
F +49 74 61 94 71 22
info@hebumedical.de
www.hebumedical.de
Seite
Page 62

Heiko Hillig
Grendelweg 5
CH-4314 Zeiningen
T +41 79 2 97 71 00
hillig@bluewin.ch
Seite
Page 180

Hissin Medizintechnik GmbH
Obere Zeil 6-8
D-61440 Oberursel/TS
T +49 61 71 2 06 25 20
F +49 61 71 2 06 25 25
info@hissin.de
www.hissin.de
Seite
Page 64

Hudson Vandam LLC
333 Hudson Street, Suite 304
New York, NY 10013
T +1 64 63 30 45 95
F +1 64 66 88 28 85
info@hudsonvandam.com
www.hudsonvandam.com
Seite
Page 144

I

IC Medical Vertriebsgesellschaft VG
Rosenstraße 42
D-73650 Winterbach
T +49 71 18 70 60 15
F +49 71 18 70 60 99
thomas.blumenschein@icmedical.de
www.icmedical.de
Seite
Page 50, 52

Idea Design Team
Lindauer Straße 46
D-88239 Wangen
T +49 75 22 97 75 80
F +49 75 22 9 77 58 20
kontakt@idea-design-team.de
www.idea-design-team.de
Seite
Page 56

Inflate UK
1 Helmsley Place
London E8 3SB UK
T + 44 20 72 49 30 34
F + 44 20 72 75 92 23
info@inflate.co.uk
www.inflate.co.uk
Seite
Page 202

Werksdesign
In-house design
Nick Crosbie
Seite
Page 202

Interbros GmbH
Aiterfeld 1
D-79677 Schönau
T +49 76 73 88 50
F +49 76 73 88 52 99
info@interbros.de
www.interbros.de
Seite
Page 48

Intuity Media Lab GmbH
Feuerseeplatz 14
D-70176 Stuttgart
T +49 7 11 94 54 55 50
F +49 7 11 94 54 55 51
mail@intuity.de
www.intuity.de
Seite
Page 186 ff

IPDD GmbH & Co. KG
Calwer Straße 11
D-70173 Stuttgart
T +49 7 11 3 26 54 60
F +49 7 11 3 26 54 61
info@ipdd.com
www.ipdd.com
Seite
Page 182

J

Just Mobile
Hähnelstraße 13
D-12159 Berlin
T +49 30 85 97 62 31
F +49 30 85 97 62 31
nils@just-mobile.eu
www.just-mobile.eu
Seite
Page 141, 142

K

Keilbach design and products
Friedhofstraße 14
D-74677 Dörzbach
T +49 79 37 99 00 77
F +49 79 37 99 00 79
design@keilbach.com
www.keilbach.com
Seite
Page 160

Werksdesign
In-house design
Peter Keilbach
mail@keilbach.com
Seite
Page 160

Stephanie Knödler
Adelmannweg 6
D-73579 Schechingen
M +49 1 76 22 50 93 95
stephanie.knoedler@gmx.de
Seite
Page 240

Kurz Kurz Design
Engelsberg 44
D-42697 Solingen
T +49 2 12 33 69 83
F +49 2 12 33 71 98
info@kurz-kurz-design.de
www.kurz-kurz-design.de
Seite
Page 90

L

C. Josef Lamy GmbH
Grenzhöfer Weg 32
D-69123 Heidelberg
T +49 62 21 84 31 47
F +49 62 21 84 32 64
fh@lamy.de
www.lamy.com
Seite
Page 182, 184

LNT Automation GmbH
Hans-Paul-Kaysser-Straße 1
D-71397 Leutenbach-Nellmersbach
T +49 71 95 58 88 90
F +49 71 95 5 88 89 28
HanspeterBurger@lnt-automation.de
www.lnt-automation.de
Seite
Page 50

M

Merida & Centurion Germany GmbH
Blumenstraße 49-51
D-71106 Magstadt
T +49 71 59 9 45 96 00
F +49 71 59 9 45 95 00
simon.oppold@merida-centurion.com
www.centurion.de
Seite
Page 156

Metz und Kindler Produktdesign
Frankfurter Straße 44
D-64293 Darmstadt
T +49 61 51 29 36 41
F +49 61 51 29 52 52
public@metz-kindler.de
www.metz-kindler.de
Seite
Page 77

L. Michow & Sohn GmbH
Wandsbeker Allee 19
D-22041 Hamburg
T +49 40 68 94 29 11
F +49 40 68 77 38
lm@michow.com
www.michow.com
Seite
Page 158

Miele & Cie. KG
Carl-Miele-Straße 29
D-33332 Gütersloh
T +49 52 41 89 41 91
F +49 52 41 89 41 40
andreas.enslin@miele.de
www.miele.de
Seite
Page 72

Minox GmbH
Walter-Zapp-Straße 4
D-35578 Wetzlar
T +49 64 41 91 76 13
F +49 64 41 91 76 12
andrea.schmidt-gel@minox.com
www.minox.com
Seite
Page 136

Modus Product Design GmbH
Alexander Benn, Harry Eppinger
Markgröninger Straße 61/1
D-70435 Stuttgart
T +49 7 11 8 70 21 05
F +49 7 11 8 70 21 06
benn@modusdesign.de
Seite
Page 36

Marko Müller
Hohlweg 8
D-54316 Hockweiler
M +49 1 79 7 56 89 60
muellermarko@gmx.de
Seite
Page 234

N

Naef Spiele AG
Untere Brühlstraße 11
CH-4800 Zofingen
T +41 62 7 46 84 84
F +41 62 7 46 84 80
info@naefspiele.ch
www.naefspiele.ch
Seite
Page 180

Neunzig° Design
Heinrich-Otto-Straße 65
D-73240 Wendlingen
T +49 70 24 50 22 00
info@neunzig-grad.com
www.neunzig-grad.com
Seite
Page 98, 106

Nimbus Group GmbH
Sieglestraße 41
D-70469 Stuttgart
T +49 7 11 6 33 01 40
F +49 7 11 63 30 14 14
info@nimbus-group.com
www.nimbus-group.com
Seite
Page 124

O

Olbrish Produkt GmbH
Bergmannstraße 68
D-10961 Berlin
T +49 30 78 70 46 66
F +49 30 78 70 46 67
goebel@olbrish.de
www.olbrish.de
Seite
Page 100 ff

Werksdesign
In-house design
Wolfgang Olbrish
olbrish@olbrish.de
Seite
Page 100 ff

P

Phocos AG
Magirus-Deutz-Straße 12
D-89077 Ulm
T +49 7 31 93 80 68 80
F +49 7 31 9 38 06 88 50
info@phocos.com
www.phocos.com
Seite
Page 130

Phoenix Design GmbH & Co. KG
Kölner Straße 16
D-70376 Stuttgart
T +49 7 11 9 55 97 60
F +49 7 11 95 59 76 99
info@phoenixdesign.com
www.phoenixdesign.com
Seite
Page 22 ff, 166, 190

Pulsmacher GmbH
Rheinlandstraße 10
D-71636 Ludwigsburg
T +49 71 41 64 86 10
F +49 71 41 6 48 61 11
dieterich@pulsmacher.de
www.pulsmacher.de
Seite
Page 202

R

Hartmut Räder Wohnzubehör GmbH & Co. KG
Kornharpener Straße 126
D-44791 Bochum
T +49 2 34 95 98 70
F +49 2 34 59 45 19
info@raeder.de
www.raeder.de
Seite
Page 78

Werksdesign
In-house design
Tobias Langner, Sarah Baumann
Seite
Page 78

Recaro Aircraft Seating GmbH & Co. KG
Daimlerstraße 21
D-74523 Schwäbisch Hall
T +49 7 91 5 03 70 00
F +49 7 91 5 03 71 63
info@recaro-as.com
www.recaro-as.com
Seite
Page 208 ff

RED GmbH
Wilhelmstraße 5a
D-70182 Stuttgart
T +49 7 11 13 78 63 13
F +49 7 11 13 78 63 29
ital@redproducts.de
www.redproducts.de
Seite
Page 54

Gerhard Reichert
Lindenweg 5
D-88719 Stetten
T +49 75 32 80 80 13
F +49 75 32 80 80 14
gerhard@reichert-design.net
www.reichert-design.net
Seite
Page 94

Rudolf Riester GmbH
Bruckstraße 31
D-72417 Jungingen
T +49 74 77 92 70 43
F +49 74 77 92 70 70
kleiner@riester.de
www.riester.de
Seite
Page 54

RUD Ketten
Rieger & Dietz GmbH & Co. KG
Friedensinsel
D-73432 Aalen
T +49 73 61 50 40
F +49 73 61 5 04 18 02
sven.cravotta@rud.com
www.rud.com
Seite
Page 214

Axel Ruhland Produktgestaltung
Teckstraße 56
D-70190 Stuttgart
T +49 7 11 6 36 86 50
F +49 7 11 6 36 86 49
mail@axelruhland.de
www.axelruhland.de
Seite
Page 32

S

Johannes Scherr
Schottenfeldstraße 73/4
A-1070 Wien
T +43 1 90 4 38 13
office@johannes-scherr.com
www.johannes-scherr.com
Seite
Page 96

Wilhelm Schimmel Pianofortefabrik GmbH
Friedrich-Seele-Straße 20
D-38122 Braunschweig
T +49 5 31 8 01 81 14
F +49 5 31 8 01 81 17
marius.hebig@schimmel-piano.de
www.schimmel-piano.de
Seite
Page 118

Werksdesign
In-house design
Sven Schneider
Seite
Page 118

Schmitz Visuelle Kommunikation
Hans Günter Schmitz
Bembergstraße 20
D-42103 Wuppertal
T +49 2 02 37 16 30
F +49 2 02 3 71 63 14
email@hgschmitz.de
www.hgschmitz.de
Seite
Page 193

Schoeller Textil AG
Bahnhofstraße 17
CH-9475 Sevelen
T +41 81 7 86 08 00
F +41 81 7 86 08 10
info@schoeller-textiles.com
www.schoeller-textiles.com
Seite
Page 179

Scopetime GmbH
Krausestraße 102
D-22049 Hamburg
T +49 40 36 16 36 16
F +49 40 36 16 36 18
info@neolog.de
www.neolog.de
Seite
Page 104

Serien Raumleuchten
Hainhäuser Straße 3-7
D-63110 Rodgau
T +49 61 06 6 90 90
F +49 61 06 69 09 22
serien@serien.com
www.serien.com
Seite
Page 126

Sick AG
Erwin-Sick-Straße 1
D-79183 Waldkirch
T +49 76 81 2 02 35 05
F +49 76 81 2 02 36 00
josef.baak@sick.de
www.sick.com
Seite
Page 30

Sick Vertriebs-GmbH
Willstätterstraße 30
D-40549 Düsseldorf
T +49 2 11 5 30 13 01
F +49 2 11 5 30 13 02
kundenservice@sick.de
www.sick.com
Seite
Page 30

SieMatic Möbelwerke GmbH & Co. KG
August-Siekmann-Straße 1-5
D-32584 Löhne
T +49 57 32 6 74 20
k.fricke@siematic.de
www.siematic.de
Seite
Page 74

Konstantin Slawinski
Housewarming objects
Wilhelm-Mauser-Straße 49 c
D-50827 Köln
T +49 2 21 28 55 39 00
F +49 2 21 28 55 39 09
hello@konstantinslawinski.com
www.konstantinslawinski.com
Seite
Page 112, 116

Slogdesign
Jürgen Hinderhofer
Waldseer Straße 13
D-88400 Biberach
T +49 73 51 1 79 33
F +49 73 51 1 79 37
hinderhofer@slogdesign.de
www.slogdesign.de
Seite
Page 156

Spek Design GbR
Schopenhauerstraße 39
D-70565 Stuttgart
T +49 7 11 74 54 31 36
F +49 7 11 74 54 31 40
p.sauter@spek-design.de
www.spek-design.de
Seite
Page 28

Speziell®
Luisenstraße 30b
D-63067 Offenbach
T +49 69 80 08 54 30
F +49 69 80 08 54 31
jp@speziell.net
www.speziell.net
Seite
Page 74

Starmed GmbH
Im Lehrer Feld 30
D-89081 Ulm
T +49 7 31 1 40 03 30
F +49 7 31 14 00 33 99
info@starmed.ulm.eu
www.starmed.eu
Seite
Page 64

Werksdesign
In-house design
René Stern
Seite
Page 64

Steng Licht AG
Hedelfinger Straße 103
D-70327 Stuttgart
T +49 7 11 23 88 00
F +49 7 11 23 88 00
mail@steng.de
www.steng.de
Seite
Page 120 ff

Werksdesign
In-house design
Andreas Steng
Seite
Page 120 ff

Martin Storz Design
Schauinslandstraße 1
D-78532 Tuttlingen
T +49 74 62 9 14 85
F +49 74 62 9 14 87
mail@storz-design.de
www.storz-design.de
Seite
Page 62

Stotz-Design.com GmbH & Co. KG
Besenbruchstraße 6
D-42285 Wuppertal
T +49 2 02 30 06 67
F +49 2 02 30 06 68
info@stotz-design.com
www.stotz-design.com
Seite
Page 48

Südmo Components GmbH
Industriestraße 7
D-73469 Riesbürg
T +49 90 81 80 30
F +49 90 81 80 31 58
thomas.feldmeier@suedmo.de
www.suedmo.de
Seite
Page 36

Supernova
Industriestraße 26
D-79194 Gundelfingen
T +49 7 61 45 87 82 10
F +49 7 61 45 87 82 19
info@supernova-lights.com
www.supernova-lights.com
Seite
Page 150 ff, 155

Werksdesign
In-house design
Marcus Wallmeyer
T +49 7 61 45 87 82 11
marcus.wallmeyer@
supernova-lights.com
Seite
Page 150 ff, 155

Synapsis Design GmbH
Teckstraße 56
D-70190 Stuttgart
T +49 7 11 2 62 11 31
F +49 7 11 2 62 26 70
mail@synapsisdesign.com
www.synapsisdesign.com
Seite
Page 30

T

Teams Design GmbH
Kollwitzstraße 1
D-73728 Esslingen
T +49 7 11 3 51 76 50
F +49 7 11 35 17 65 25
info@teams-design.de
www.teams-design.com
Seite
Page 34

Hadi Teherani
Kaiserkai 26
D-20457 Hamburg
T +49 40 4 44 05 43
info@haditeherani.de
www.haditeherani.de
Seite
Page 206

Uwe Thamm
Hauptstraße 48
D-74595 Langenburg
T +49 79 05 51 22
F +49 79 05 52 01
info@thamm-design.com
www.thamm-design.com
Seite
Page 212

Tools Design
Henrik Holbæk, Claus Jensen
Rentemestervej 23A
DK-2400 Kopenhagen NV
T +45 22 44 41 14
F +45 38 19 41 13
email@toolsdesign.dk
www.toolsdesign.dk
Seite
Page 80, 82, 108, 141, 142

Samuel Treindl
Nieberdingerstraße 26
D-48155 Münster
T +49 1 76 82 15 95 04
mail@samuel-treindl.de
www.samuel-treindl.de
Seite
Page 116

TRI International
Steinstraße 28
D-88339 Bad Waldsee
T +49 75 24 70 33 26
F +49 75 24 70 34 74
j.wolber@tri-international.de
www.tri-international.de
Seite
Page 94

V

Vaude Sport GmbH & Co. KG
Vaude-Straße 2
D-88069 Tettnang
T +49 75 42 5 30 60
F +49 75 42 53 06 60
info@vaude.com
www.vaude.com
Seite
Page 168 ff, 172, 174

Werksdesign
In-house design
Hans-Thomas Langowski
Seite
Page 168 ff

Werksdesign
In-house design
Matthias Kimmerle,
Hans-Thomas Langowski
Seite
Page 172

Werksdesign
In-house design
Katrin Bojarski, Philipp Ziegler
Seite
Page 174

Viessmann Werke GmbH & Co. KG
Viessmannstraße 1
D-35108 Allendorf
T +49 64 52 7 00
F +49 64 52 70 27 80
info@viessmann.com
www.viessmann.com
Seite
Page 22 ff, 190

Volkswagen Design
Volkswagen AG
Brieffach 1669/6
D-38436 Wolfsburg
T +49 53 61 92 00 68
www.volkswagen.de
Seite
Page 136

W

Wagner System
Tullastraße 19
D-77933 Lahr
T +49 7821 9 47 70
F +49 7821 9 47 60
info@wagner-system.de
www.wagner-system.de
Seite
Page 148

Namenregister
Index of names

Die fett gedruckten Seitenzahlen verweisen auf den Katalogteil, die mager gedruckten führen zum Adressenverzeichnis.

Page numbers in bold type refer to the catalogue section, while those in lightface type refer to the index of addresses.

Fotonachweis
Photographs
(acknowledgements)

soweit von den Einreichern erwünscht:

information provided by competitors only:

Seite
Page 18
Mathias Ritzmann, Burg Halle

Umschlag
Cover

Robert Westrich
Er lebt und arbeitet als freischaffender Fotograf in Stuttgart.
www.robertwestrich.com

Robert Westrich
He lives and works as a freelance photographer in Stuttgart.
www.robertwestrich.com